Models and Modeling

Models and Modeling

An Introduction for Earth and Environmental Scientists

Jerry P. Fairley

WILEY Blackwell

This edition first published 2017 © 2017 by John Wiley & Sons, Ltd

Registered Office
John Wiley & Sons, Ltd, The Atrium, Southern Gate, Chichester, West Sussex, PO19 8SQ, UK

Editorial Offices
9600 Garsington Road, Oxford, OX4 2DQ, UK
The Atrium, Southern Gate, Chichester, West Sussex, PO19 8SQ, UK
111 River Street, Hoboken, NJ 07030-5774, USA

For details of our global editorial offices, for customer services and for information about how to apply for permission to reuse the copyright material in this book please see our website at www.wiley.com/wiley-blackwell.

Library of Congress Cataloging-in-Publication Data

Names: Fairley, Jerry P., author.
Title: Models and modeling : an introduction for earth and environmental scientists / Jerry P. Fairley.
Description: Chichester, UK ; Hoboken, NJ : John Wiley & Sons, 2016. |
   Includes bibliographical references and index.
Identifiers: LCCN 2016024807 (print) | LCCN 2016033087 (ebook) | ISBN 9781119130369 (pbk.) |
   ISBN 9781119130383 (pdf) | ISBN 9781119130376 (epub)
Subjects: LCSH: Earth sciences–Mathematical models. | Environmental sciences–Mathematical models.
Classification: LCC QE33.2.M3 F35 2016 (print) | LCC QE33.2.M3 (ebook) | DDC 550.1/5118–dc23
LC record available at https://lccn.loc.gov/2016024807

A catalogue record for this book is available from the British Library.

Wiley also publishes its books in a variety of electronic formats. Some content that appears in print may not be available in electronic books.

Cover image: iStockphoto/Alexander Shirokov

Set in 9.5/13pt Meridien by SPi Global, Pondicherry, India

Printed in Singapore by C.O.S. Printers Pte Ltd

10 9 8 7 6 5 4 3 2 1

To my father, for teaching me to do math one step at a time.

Contents

About the companion website, xi

Introduction, 1

1 Modeling basics, 4
    1.1   Learning to model, 4
    1.2   Three cardinal rules of modeling, 5
    1.3   How can I evaluate my model?, 7
    1.4   Conclusions, 8

2 A model of exponential decay, 9
    2.1   Exponential decay, 9
    2.2   The Bandurraga Basin, Idaho, 10
    2.3   Getting organized, 10
    2.4   Nondimensionalization, 17
    2.5   Solving for θ, 19
    2.6   Calibrating the model to the data, 21
    2.7   Extending the model, 23
    2.8   A numerical solution for exponential decay, 26
    2.9   Conclusions, 28
    2.10 Problems, 29

3 A model of water quality, 31
    3.1   Oases in the desert, 31
    3.2   Understanding the problem, 32
    3.3   Model development, 32
    3.4   Evaluating the model, 37
    3.5   Applying the model, 38
    3.6   Conclusions, 39
    3.7   Problems, 40

4 The Laplace equation, 42
    4.1   Laplace's equation, 42
    4.2   The Elysian Fields, 43
    4.3   Model development, 44
    4.4   Quantifying the conceptual model, 47
    4.5   Nondimensionalization, 48
    4.6   Solving the governing equation, 49
    4.7   What does it mean?, 50

4.8   Numerical approximation of the second derivative, 54

4.9   Conclusions, 57

4.10   Problems, 58

5 The Poisson equation, 62

5.1   Poisson's equation, 62

5.2   Alcatraz island, 63

5.3   Understanding the problem, 65

5.4   Quantifying the conceptual model, 74

5.5   Nondimensionalization, 76

5.6   Seeking a solution, 79

5.7   An alternative nondimensionalization, 82

5.8   Conclusions, 84

5.9   Problems, 85

6 The transient diffusion equation, 87

6.1   The diffusion equation, 87

6.2   The Twelve Labors of Hercules, 88

6.3   The Augean Stables, 90

6.4   Carrying out the plan, 92

6.5   An analytical solution, 100

6.6   Evaluating the solution, 109

6.7   Transient finite differences, 114

6.8   Conclusions, 118

6.9   Problems, 119

7 The Theis equation, 122

7.1   The Knight of the Sorrowful Figure, 122

7.2   Statement of the problem, 124

7.3   The governing equation, 125

7.4   Boundary conditions, 127

7.5   Nondimensionalization, 128

7.6   Solving the governing equation, 132

7.7   Theis and the "well function", 134

7.8   Back to the beginning, 135

7.9   Violating the model assumptions, 138

7.10   Conclusions, 139

7.11   Problems, 140

8 The transport equation, 141

8.1   The advection–dispersion equation, 141

8.2   The problem child, 143

8.3   The Augean Stables, revisited, 144

8.4   Defining the problem, 144

8.5   The governing equation, 146

8.6   Nondimensionalization, 148

8.7   Analytical solutions, 152

8.8   Cauchy conditions, 165

8.9   Retardation and dispersion, 167

8.10   Numerical solution of the ADE, 169

8.11   Conclusions, 173

8.12   Problems, 174

9 Heterogeneity and anisotropy, 177

9.1   Understanding the problem, 177

9.2   Heterogeneity and the representative elemental volume, 179

9.3   Heterogeneity and effective properties, 180

9.4   Anisotropy in porous media, 187

9.5   Layered media, 188

9.6   Numerical simulation, 189

9.7   Some additional considerations, 191

9.8   Conclusions, 192

9.9   Problems, 192

10 Approximation, error, and sensitivity, 195

10.1   Things we almost know, 195

10.2   Approximation using derivatives, 196

10.3   Improving our estimates, 197

10.4   Bounding errors, 199

10.5   Model sensitivity, 201

10.6   Conclusions, 206

10.7   Problems, 207

11 A case study, 210

11.1   The Borax Lake Hot Springs, 210

11.2   Study motivation and conceptual model, 212

11.3   Defining the conceptual model, 213

11.4   Model development, 215

11.5   Evaluating the solution, 224

11.6   Conclusions, 229

11.7   Problems, 230

12 Closing remarks, 233

12.1   Some final thoughts, 233

Appendix A A heuristic approach to nondimensionalization, 236

Appendix B Evaluating implicit equations, 238

B.1   Trial and error, 239

B.2   The graphical method, 239

B.3   Iteration, 240

B.4   Newton's method, 241

Appendix C Matrix solution for implicit algorithms, 243

    C.1   Solution of 1D equations, 243

    C.2   Solution for higher dimensional problems, 244

    C.3   The tridiagonal matrix routine TDMA, 244

Index, 247

About the companion website

This book is accompanied by a companion website:

www.wiley.com/go/Fairley/Models

This website includes:

- Powerpoints of all figures from the book for downloading
- Solutions to many of the problems given in the chapters

Introduction

Before serious discussion on any topic can take place, it is usually necessary to define one's terms. In the present instance, we need to ask the following question: "what, exactly, do we mean when we speak of a 'model'?" A reasonable definition of the term might be *a simplified or idealized representation of reality.* Although this is a pretty broad definition, it is perhaps the only one that really embraces everything we refer to as "a model."

In order to better understand what is, and what is not, a model, we can examine some possible examples. Which of the following would be considered models?

- A map
- A model train
- A model United Nations
- A fashion model
- An atom

Of course, by our definition all of these things would qualify as models. Students will sometimes argue that "an atom" is a physical object and, therefore, not a model. However, everything we know about atoms comes from observing their effects, and our understanding and depiction of atoms has changed and evolved over time. In fact, our current understanding of atoms and subatomic particles is so far outside our everyday experience that our descriptions are entirely probabilistic. As a result, I maintain that any discussion of "atoms" is necessarily a discussion of models—simplifications and idealizations of things that are beyond our direct knowledge.

In this book, we will primarily concentrate on a subset of models known as *mathematical models.* A mathematical model is a kind of model that is formulated in terms of mathematical concepts such as constants, variables, functions, derivatives, and so on. The nominal goal of a mathematical model is to give quantitative indications of the behavior of some aspect of a system.

Note that I said we would *primarily* be concentrating on mathematical models. In fact, the word "primarily" glosses over an important fact about the development of mathematical models: *every mathematical model is the quantification*

Models and Modeling: An Introduction for Earth and Environmental Scientists, First Edition. Jerry P. Fairley.
© 2017 John Wiley & Sons, Ltd. Published 2017 by John Wiley & Sons, Ltd.
Companion website: www.wiley.com/go/Fairley/Models

of an underlying conceptual model; that is, *a mathematical model is the quantification of our idea of how a system works*. This is a very important, but often overlooked, aspect of model development. The fact is that model development requires many decisions to be made about which aspects of the system are important, and which aspects may be neglected. If the decisions are poor, a poor model will result. If the decisions are good, a good model *may* result (no guarantee). In any case, before a *mathematical* model can be developed, it is first necessary to define a *conceptual model*, or conceptual understanding of the system. The usefulness of the quantitative model, and the reliability of its representation of the system, will be wholly dependent on this conceptual understanding.

Another aspect of models that is deserving of discussion is the uses to which models may be put. A sample of these might be the following:

- To teach
- To "predict" system behavior or the consequences of actions
- To develop or test theories
- To plan or design field/laboratory tests
- As propaganda to support a partisan viewpoint

Of these, the use of models as propaganda is probably the most common application for models (e.g., think about the usefulness of fashion models and model cars in selling products). It is tempting to think that "scientific" (i.e., mathematical) models are exempt from issues of partisanship, but this is far from the truth. Because every model is a quantification of an individual's conceptual understanding, all models are developed from a partisan point of view; that is, *every model contains bias!* Therefore, the first step in evaluating any model is to understand the motivations and biases of the developers.

Another topic it would be appropriate to touch on is *model fallibility*. It is not uncommon to hear statements similar to "it must be true, because the model says so," and models are commonly depicted in popular culture as unfailing guides to the future. In reality, the knowledgeable modeler understands that a model embodies our conceptual understanding of some system, and is therefore incomplete and prone to error. Furthermore, there are an infinite number of alternative models that could be devised to represent any given system. Since it is impossible to test all the alternatives for a superior model, *a model can never be shown to be the "true" representation of a system*.

It is also easy to make the mistake of thinking that a model is reliable because it has been "validated" or "verified." In my opinion, the idea of "model validation" is misleading at best; it implies that, because a model has produced good results in the past, future predictions may be taken uncritically. In fact, every new prediction or new situation to which a model is applied will take a model into scenarios outside of the conditions for which it was calibrated and tested. Under these circumstances, the modeler must always be alert to the possibility that new processes, changes in parameters, or insufficiencies of the underlying conceptual model will cause the model to fail.

Instead of using the terms "validated" or "verified," I prefer to think in terms of *confidence building*. If a model performs well when tested against observations, we gain confidence in the model's output, and each additional test increases our confidence. We retain our confidence as long as the model continues to perform reliably; however, we never lose sight of the fact that, at some time in the future, the model parameterization or underlying conceptualization may be insufficient for the task at hand, and the model may fail. Keep in mind that, although a model can never be proven to be "right," it only takes one counterexample to prove it wrong.

Usually, models are created for some useful purpose—to understand some phenomenon, plan some test, or make some prediction—and I am a believer in the idea that models should be useful. However, I would be less than honest if I didn't say that one of the reasons for developing models was the pleasure to be had in their development. I believe this is as good a reason to develop a model as any, and perhaps better than many others. As you work through the models presented in the following chapters, I hope that a little of the enjoyment and gratification that comes from working with models will shine through.

CHAPTER 1

Modeling basics

Chapter summary

Regardless of whether the subject to be modeled is a groundwater flow system, the growth of a bacterial colony, the flight of a projectile shot from a cannon, or any other system, the construction of a model is not a simple step-by-step procedure. Because the modeler must make decisions about which processes should be included in the model and which will be neglected, what domain will be used, and so on, the development of a model is as much art as science. Notwithstanding the creative aspects of model development, there are some simple but important rules that should be followed. In this section, we will examine three basic rules for model development, briefly discuss some important aspects of model formulation, and make some suggestions for evaluating a model's performance. These rules and suggestions will form the basis for all of the examples of model development in the following chapters.

1.1 Learning to model

How to develop a model of a physical system, when to believe and when not to believe the model output, and how to determine whether the model predictions have any relevance to real life are common and central questions that must be answered by the would-be modeler every time a new situation is encountered. Most commonly, modelers are shown how to use a software package (or asked to read the documentation for a software package), and then assumed to be sufficiently competent to produce reliable predictions of system behavior. Even a moment's reflection will show that this is a nonsensical way to go about learning the craft of modeling, and this attitude has been largely responsible for the proliferation of bad models and the subsequent lack of confidence in modeling (and modelers).

Ideally, a modeler would gain experience and a deep understanding of the process of model development while working as an "apprentice" under a skilled modeler. This desirable state of affairs is seldom met with in the real world, however. The purpose of this book is to provide some guidance for those aspiring

Models and Modeling: An Introduction for Earth and Environmental Scientists, First Edition. Jerry P. Fairley.
© 2017 John Wiley & Sons, Ltd. Published 2017 by John Wiley & Sons, Ltd.
Companion website: www.wiley.com/go/Fairley/Models

modelers who do not have the advantage of serving such an apprenticeship. Although not a substitute for the teaching and advice of an experienced modeler, it is to be hoped that the rules and examples in this and the following sections will at least keep the novice modeler from falling into some of the more obvious pitfalls associated with the mathematical modeling of physical systems.

1.2 Three cardinal rules of modeling

It is probable that, over time, many hundreds of "rules" have been made up regarding the construction, evaluation, and application of mathematical models. Most of these purported rules would be better classified as "suggestions," "considerations," or even, in some cases, as "superstitions." Over many years of making and using models, however, I have become convinced that the following three cardinal rules should be followed at all times in the development of a mathematical model:

1. Always know exactly the objective of model development.
2. The model you develop should be appropriate for the available data.
3. Start with the simplest possible model of the system, even if it is completely unrealistic. Once you thoroughly understand this preliminary model, add complexities to the model *one at a time* until you arrive at a satisfactory representation of the system.

In my experience, all three of these rules are routinely overlooked by modelers, and many poor and inappropriate models have resulted. We will briefly consider each of these rules here, but, more importantly, they are bound into the fiber of every model developed in the following chapters.

1.2.1 Rule 1: Know your model objective

It is common for a modeler to start a modeling investigation with the objective of "making a model of the aquifer" or with a similarly vague idea of what is to be accomplished. I cannot state strongly enough that a modeler must know exactly what s/he is trying to accomplish before ever putting pen to paper (or typing an input parameter). The more precisely the objectives of the model are known, the more likely the investigation will be successful. Make a habit of writing down the objective of your model, and be ready and willing to reduce your objective to a single sentence. The objective "to model the wells in the Grande Ronde Aquifer" is a very poor statement of purpose; a better (although still insufficient) objective is "to determine the influence of pumping well MW-4 on nearby wells." Better yet (and possibly sufficient to begin an investigation) is "to estimate the change in head in wells MW-1 and MW-3 that results from a 24-hour constant rate pump test in well MW-4." The examples in the following chapters always include a statement of that which is to be found; hopefully, after working through the examples, the reader will have a clear idea of how to formulate model objectives.

Although the temptation to "just get modeling" and "show some results" may be strong, you will always be better off if you first make certain you understand exactly what it is you want to achieve, and formulate a plan to reach that goal.

1.2.2 Rule 2: Make your model appropriate for your data

Hydrogeologists and environmental scientists are often working in data-poor environments. There is usually little to be gained from building a three-dimensional (3D), coupled saturated–unsaturated zone model with heterogeneous property sets when the only data to constrain the model come from a single aquifer test. Often in these situations a simple analytical model will give results that are as reliable as (or more reliable than) a complex numerical simulation. Furthermore, complex numerical simulations are often misleading, since it is tempting to think that, because they are complicated, they are realistic. What non-modelers (and many modelers) are unaware of is the fact that any computer model is solving the same equations that the modeler can write down with a pencil and paper. If there are few data to support the added complexity in terms of spatially varying properties, time-dependent recharge or boundary conditions, and so on, then the complicated numerical simulation may in fact be a worse representation of the system than a greatly simplified analytical model.

It should be said that many clients, regulatory agencies, and other down-stream users of model output will push for complex numerical simulations in spite of the paucity of data to support such simulations. Although economic, political, or regulatory pressures may force a modeler to undertake the development of 3D simulations when only 1D simulations are justified, or a transient model when a steady-state model would do, the modeler should at all times be aware of the limitations of the models s/he is working with. By following Rule 3 (Section 1.2.3), the savvy modeler will be able to develop the more complex model demanded by the client while still maintaining her or his integrity and a high standard of modeling ethics.

1.2.3 Rule 3: Start simple and build complexity

When faced with a complex and difficult real-life situation, it is tempting to start out by developing a model that includes the most important processes. For example, if the goal is to understand the impact of a pumping well on other nearby wells, a novice modeler might want to build a model that includes variations in the rate of pumping, recharge from rainfall or snowmelt, the influence of changing water levels in a nearby lake, and other similar items that are clearly needed for a realistic representation of the system. The problem is that there is no way, in such a complex conceptualization, for the modeler to determine if the model output makes sense or not. Your first attempt at modeling a system should always be the simplest possible representation. Rather than modeling a 3D transient system, begin by modeling a 1D or 2D steady-state system with no source terms or other complexities. Although this may not be a realistic representation of the system, at least the modeler will know if the results are reasonable. Next, the modeler may add a spatially and temporally constant source term; again, evaluate the results. Do they make sense? Can you convince yourself the output is reasonable? If so, add another complexity, reevaluate, and so on, until the final product is one that you both understand and believe in. Never go on to the next step until you have complete confidence in, and an intimate understanding of, the current step—as well as all the steps that lead to the current step.

1.3 How can I evaluate my model?

Particularly with regard to Rule 3 (Section 1.2.3), one may rightly ask the question: "how do I know my model is giving reasonable results?" There is no easy answer to this question; in part, knowing a model is giving reasonable results is the product of experience with models. There are, however, some simple suggestions that can help to uncover problems with a model; these are described in the following text.

1.3.1 Test model behavior in the limits

The quickest way to begin an examination of model behavior is to check the behavior of a model in the limits. What is the model behavior at time $t = 0$? What about as $t \to \infty$? If you set a parameter to 0 or check the limit as it goes to ∞, is the result what you would expect? These kinds of tests are most readily carried out with analytical models, but, with some ingenuity, they can usually be applied even to complex numerical simulations.

1.3.2 Look for behavior congruent with the governing equations

Even very complex numerical simulations are based on a few well-known equations such as the Laplace equation, Poisson's equation, and the transient diffusion equation. Each of these equations implies particular behaviors, and you should check to make certain your model is behaving in a fashion that accords with your understanding of the underlying governing equations. (We will examine the characteristic behaviors of the most common equations in the following chapters.) Check for maxima and minima, look at the curvature of the predicted model surface, and examine any discontinuities in the output. Are these features present (if you expect them) or not (if you don't expect them), and are they in the appropriate places? It is a good sign if your expectations, based on your understanding of the characteristics of the governing equations, are realized. If the behavior you observe in your model output is different than your expectations, you need to understand why these differences arise before moving forward.

1.3.3 Nondimensionalization

Whenever possible, you should nondimensionalize your model (nondimensionalization is described in Appendix A, and illustrated throughout this text). Note that, although you will rarely be able to nondimensionalize a commercial simulation package, the model is based on mathematical equations that can always be nondimensionalized. Nondimensionalization carries with it a number of benefits; in particular the following:

- Nondimensionalization reduces the governing equations to their most basic functional form, which helps to clarify the expected behavior.
- Examination of the nondimensional form of the governing equation is the easiest and most certain way to identify which parts of an equation are relevant and which parts may be neglected. In this way, simplifications of the original equations may often be made (and, equally important, justified).
- Nondimensional plots of model output are the most compact way of presenting the model results. Admittedly, your target audience may not be equipped to understand dimensionless results; in this case, it is up to the modeler to either present dimensional results or educate their audience regarding the nondimensional ones.
- The nondimensionalization process results in the identification of the controlling dimensionless parameters. These parameters control the behavior of the equation, allowing the modeler to readily identify parameter ranges over which behavioral changes will take place. Furthermore, the dimensionless parameters show the modeler which dimensional parameters can be uniquely identified and which cannot.

1.4 Conclusions

As was stated in Section 1.1, following the rules and suggestions laid out in this chapter won't guarantee success, nor will it make you a modeler (only time and experience will do that). Hopefully, however, the ideas presented here will help you avoid some of the most common traps that inexperienced modelers tend to fall into. In the following chapters, I will develop a number of models; as you follow these developments, watch for the application of these basic principles. Ask yourself how you could apply these principles to your own problems. As with any creative endeavor, there are rules and guidelines that can be applied, but the ultimate responsibility for the final product belongs with the artist.

CHAPTER 2

A model of exponential decay

Chapter summary

Possibly, the most basic building block for mathematical models of physical systems is the process of exponential growth and decay; models of exponential decay processes are encountered in every branch of science and engineering. Here, we examine the problem of modeling discharge from a high alpine basin, which leads to the representation of basin discharge as an exponential decay process. Although simple, the model is quite general and may apply to many springs and streams, as well as to the electrical charge stored in a capacitor, the temperature of a heated piece of steel quenched in cold water, and the decay of radioactive materials.

2.1 Exponential decay

Exponential decay processes are ubiquitous in nature. The essence of exponential decay is that the rate at which some quantity is lost is determined by the amount of the quantity that remains behind. We are all familiar with this process on an intuitive level; for example, a can of soda placed in the refrigerator contains quite a bit of heat (relative to the temperature of the refrigerator). The temperature gradient between the refrigerator and the soda is large at early times; thus, the soda at first cools rapidly. However, as time goes on, the difference in temperature between the soda and the refrigerator lessens; as a result, the rate of cooling also decreases. Theoretically, the soda will never actually reach the temperature of the refrigerator, but it will come arbitrarily close over a long period of time. We say that the temperature of the soda decays exponentially to the temperature of the refrigerator.

Much of the behavior observed in the natural world can be expressed in terms of differential equations—equations that show relationships between the rates of change in different quantities—and, because they possess the property of being their own derivatives, the solutions to differential equations very commonly involve exponential functions. Thus, it should be no surprise to find that exponential functions are encountered widely in natural phenomena.

Models and Modeling: An Introduction for Earth and Environmental Scientists, First Edition. Jerry P. Fairley.
© 2017 John Wiley & Sons, Ltd. Published 2017 by John Wiley & Sons, Ltd.
Companion website: www.wiley.com/go/Fairley/Models

Although exponential decay models are common to many disciplines, each discipline refers to them by a different name. In heat transfer, models of exponential decay are called "lumped capacitance" models; similarly, electrical engineers use the term "RC models" (for "resistance–capacitance"). Hydrogeologists, who use exponential decay models to represent spring flow, basin discharge, and similar processes, often refer to such models as "baseflow recession models." In the following sections, we will develop a simple exponential decay model for basin discharge, using as an example a hypothetical watershed in central Idaho.

2.2 The Bandurraga Basin, Idaho

Suppose you are approached by the US Forest Service regarding the management of water yield from an alpine basin in central Idaho. The basin receives a variable amount of precipitation in the form of rain and snow each year; in general, there is little or no precipitation after April. As one would expect, discharge in the stream that drains the basin (the "main stem") peaks about the time precipitation in the basin ceases, and stream flow decreases over time until precipitation resumes in the fall. Water from the basin is captured in a down-valley reservoir and used to operate a micro-hydroelectric generator system that provides power for the Forest Service wilderness visitor center. The water manager's problem is this: some years there is an overabundance of discharge from the basin, and the excessive runoff can damage the reservoir; in other years, the water manager drains (draws down) the reservoir in early spring, making room in the reservoir to store the basin yield, but there is not always enough water to refill the reservoir. When spring and summer flows are too low to refill the reservoir, the visitor center can experience power shortages in late summer. The water manager's question is, can you help?

2.3 Getting organized

Most modelers are familiar with this kind of request: can you help? Can you make a model of this? Whenever you receive a request of this type, the first thing you should do is work to understand the exact nature of the problem and what is expected of you (i.e., Section 1.2.1). Requests for models of physical systems are rarely well-thought-out, at least in part because few non-modelers really understand the craft of modeling. Regardless of whether or not the client is clear in her/his description of the problem and desired objectives of the model, you can be certain that you, as the modeler, will bear the brunt of the client's unhappiness if your deliverable does not meet the client's (often unstated) expectations. As a result, it is up to you to discuss the problem with the client until you clearly understand what is required; furthermore, you must share your understanding with the client, to be certain that the client knows what to expect when s/he receives your final report.

2.3.1 Observable quantities

This is probably a good point to pause and reflect on what I call "observable quantities." When formulating your model objectives, it is important to keep in mind what can be observed and what cannot. A model that makes predictions that cannot be checked is rarely useful; as a result, the quantity you select as your model output must, at least in principle, be amenable to measurement. In this example, there would be very little point to making predictions of average saturation in the basin, or even of the quantity of water stored in the basin, because there is no way to measure an average basin saturation or independently verify the amount of water in the basin at any given time. Furthermore, it is not clear how the amount of water in the basin (or the saturation) can be related to the quantity of discharge from the basin. In a sort of "worst case," the Forest Service's water manager may equate the "total water stored in the basin" with the amount of water available for power generation. This will almost certainly translate into a resource shortfall, and the modeler will bear the responsibility for such a shortfall because s/he provided misleading model predictions.

2.3.2 Stating the model objective

On the basis of the foregoing discussions (Sections 2.3.1 and 1.2.1), it is clear we must formulate our model objective with consideration of our observable quantity (in this case, the discharge from the basin) and with other expectations we have for the solution (i.e., to what use will we put the model output?). In this case, I propose a good expression for that which is to be found may be the following:

FIND: An expression for discharge from the Bandurraga Basin, Idaho, at any time following the cessation of precipitation in an annual cycle. The expression must allow the calculation of cumulative discharge between any two specified points in time (e.g., from May 1 to August 1).

Note that we have stated exactly what the model output should be (the basin discharge as a function of time) and, additionally, we have stated the expectation that we will be able to integrate the solution between any two points in time (either analytically or numerically). The first of these objectives tells us we don't necessarily need to know the spatially distributed head in the basin, the temperature of the discharge, the dissolved load in the main stem, the amount of power that can be generated from the water in the basin, or similar extraneous information. The second part of the statement of model objectives starts us thinking about what kinds of models might or might not be appropriate; for example, we probably don't want to devise a statistical model that gives the mean discharge of all previous years on day X, because such a model can't give predictions of cumulative discharge specific to the current water year (although it would certainly be possible to devise a probabilistic model for this situation).

2.3.3 What data are available?

Once we have a clear statement of our model objective, the next step is to marshal our resources in terms of how much we know about the situation. What data are available? What facts about the model domain are known to us? In this case, we will assume we have the type of information we could reasonably expect to be available for a wilderness basin under Forest Service management.

KNOWN:

- Discharge from the basin at some initial time (say May 1)
- Basin area, topography, and geometry
- Type of vegetative cover in the basin
- Twenty-year record of precipitation, discharge, and temperatures in the basin
- Basin geology (rock types, geological structures, faults, etc.)
- Information on soil types and thicknesses

It is likely that not all of this information will be useful to us, but it isn't always clear in advance what will be and what will not be useful. Furthermore, the process of thinking through what is known about a system is a crucial part of developing an overall understanding of the problem and of conceptualizing an approach to its solution.

2.3.4 What can we assume?

The next step in conceptualizing an approach to model development is to begin a list of our model assumptions. This list of assumptions will change as we formulate the model; it may grow or shrink as we relax some of our initial assumptions. Regardless of the way the list evolves over time, *you must keep careful track of your modeling assumptions!* You will need to know your model assumptions for your final report (or journal article, thesis, etc.), and it will be helpful to have them all gathered together in one place.

There are many possible assumptions that could be made at this point in the development of our model; for example, we could assume the basin comprises homogeneous and isotropic geological materials (this is a common assumption of groundwater models), or that precipitation in the basin is equal to the average of the precipitation over the past 20 years. In general, however, it is best to make a minimal number of assumptions at the start, and then to impose additional assumptions as needed during model development (while keeping careful track as new assumptions arise). In this case, we can (at least initially) make do with a very short list of assumptions.

ASSUME:

- No additional precipitation in the basin after time $t = 0$.
- We can neglect hydraulic gradients and variations on a subbasin scale, and consider the basin as a "lumped system."

In general, if you state your assumptions clearly when you report the findings of your study, you are on safe ethical ground. If you don't state your assumptions explicitly, however, you will be in a very poor position when your conclusions don't pan out. As a result, I strongly recommend that you keep a careful and accurate list of your model assumptions, and keep it up to date.

2.3.5 Finding an approach

At this point, we have reached one of those critical junctures that appears in the formulation of every model: we must settle on a conceptual model of our system. How does the system work? Our mathematical model will be a quantification of our conceptualization of the system, so the success of our endeavor depends on how well we understand the system and on the appropriateness of our abstraction of the system. Our conceptualization must be simple enough to be tractable (the simpler, the better, in terms of time and money). At the same time, it must include sufficient detail to provide a realistic representation of the system (i.e., to represent the system accurately enough to allow us to achieve our objectives).

Given our need to develop a realistic conceptualization of the system, it is tempting to rush off to construct some kind of numerical model of the basin that includes evaporation, transpiration, recharge from rainfall and snowmelt, and so on. The question we must ask, though, is, *what model would be appropriate, given the level of information we have about the system and the objectives of our model?*

As it turns out, a more complex model is rarely a better model. William of Ockham is generally credited with formulating the guiding principle of model development in the 1300s.[1] His statement advocating parsimony and elegance in reasoning, *Numquam ponenda est pluralitas sine necessitate* ("Plurality is never to be posited without necessity") (Badius and Trechsel, 1495), has come to be known as "Ockham's Razor." These days, this principle is usually stated as follows:

Ockham's Razor: When choosing between two explanations for a phenomenon, the simplest explanation is most likely to be the best, all other things being equal.

Of course, if a simple model neglects important processes or fails to produce output that adequately reproduces the observations, a more complex model may be warranted. Following the cardinal rules of modeling, however (Rules 2 and 3, Section 1.2), suggests that you should first seek a simple model that is appropriate to the existing data, and add complexity only if it is needed. In other words, you should remember William of Ockham's principle and apply it to all your model development problems.

For the problem at hand, a little thought will show that a simple analytical model is likely to serve us better than a complex numerical simulation. In part, this is because of the time that is required to set up a numerical simulation, and in part because we have few data to constrain the input parameters of such a model. Furthermore, a numerical model of the basin—if developed correctly and with sufficient data—may yield a great deal of information about the distribution of hydraulic head in the basin, the evolution of soil moisture content over time, and so on, but these quantities are completely irrelevant to our model objectives. In this case, the time and expense of developing a numerical simulation of the basin are not justified based on our objectives and the existing data. At the very least, if we develop a simple representation of the system, we can get an overall idea of how the basin works, and then reevaluate our need for a more complex model.

On the basis of these considerations, we can propose the following approach to modeling discharge from the Bandurraga Basin:

APPROACH: Conceptualize the basin as a large tank, discharging water from an outlet in the bottom of the tank to represent streamflow draining from the basin into the main stem. Use mass balance to formulate a relationship for the discharge of water from the tank as a function of time.

This simple formulation is known as a "lumped system" model, because most of the processes affecting discharge are dealt with in aggregate via a group of effective parameters. Drawing a cartoon of the conceptual model and labeling the salient quantities (Figure 2.1) will help to solidify the important points of the model, as well as assisting in developing a notation for the problem. In the

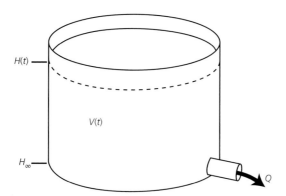

Figure 2.1 Conceptual representation of a lumped capacitance (baseflow recession) model of the Bandurraga Basin.

following sections, we will see how this lumped system model applies to the problem of modeling discharge from our hypothetical basin.

2.3.6 Executing the plan

The principle of mass balance (or energy balance for heat transfer models) underlies all models in this book, and the majority of deterministic groundwater flow models in use today. Simply stated, mass balance requires the mass coming into our model domain (M_{in}) to equal the outgoing mass (M_{out}), with any differences between those two quantities reflected in a change of mass stored in the model domain (ΔM):

$$M_{in} = M_{out} + \Delta M. \tag{2.1}$$

In the present instance, we have assumed that no mass is entering the model domain after time $t = 0$; therefore, $M_{in} = 0$ and we can rewrite Equation 2.1 as:

$$\Delta M = -M_{out}. \tag{2.2}$$

We now need to find expressions for each of the terms in Equation 2.2. The M_{out} term can be written as:

$$M_{out} = Q\rho\Delta t, \tag{2.3}$$

where ρ is the density of water [M/L^3], Q [L^3/T] is the discharge, and Δt is an arbitrary, but small, unit of time. Similarly, ΔM can be written:

$$\Delta M = \Delta V\rho, \tag{2.4}$$

where ΔV is the change in the volume of water stored in the basin [L^3]. Checking the right-hand sides of Equations 2.3 and 2.4 confirms that the units are those of mass, as desired. Substituting Equations 2.3 and 2.4 into Equation 2.2 yields

$$\Delta V\rho = -Q\rho\Delta t. \tag{2.5}$$

We can cancel the density terms[2] in Equation 2.5 and divide through by Δt, which gives

$$\frac{\Delta V}{\Delta t} = -Q. \tag{2.6}$$

To proceed further, we need to expand the expressions on both sides of Equation 2.6. We can apply a linearized form of Darcy's law

$$Q = -KA\frac{H_\infty - H(t)}{L_c} \tag{2.7}$$

to expand the expression for discharge (Q) on the right-hand side of the equation. In Equation 2.7, K is the effective hydraulic conductivity [L/T] of the basin, A is the effective area [L^2], L_c is the characteristic length of the basin [L], and H is the hydraulic head in the basin [L] (the subscript ∞ indicates the head at which the basin is considered "completely drained," see Figure 2.1). In addition, we can expand the left-hand side of Equation 2.6 using the chain rule

$$\frac{\Delta V}{\Delta t} = \frac{\Delta V}{\Delta H}\frac{\Delta H}{\Delta t}. \tag{2.8}$$

This strategy for changing variables is a common application of the chain rule in applied analysis.

In words, Equation 2.8 says that the change in the volume of water stored in the aquifer over an arbitrary time ($\Delta V/\Delta t$) is equal to the volume of water released from storage given a unit change in head ($\Delta V/\Delta H$) times the change in head over time ($\Delta H/\Delta t$). In particular, the $\Delta V/\Delta H$ term is a description of the total volume of water released from the entire basin if the head in the basin drops by a unit amount. A related quantity is the storativity (S), which is the volume of water released from storage by a unit drop in head per unit area of aquifer (storativity is unitless). As a result, we can rewrite the $\Delta V/\Delta H$ term as SA; Equation 2.8 can then be written:

$$SA\frac{\Delta H}{\Delta t} = -KA\frac{H(t) - H_\infty}{L_c}. \tag{2.9}$$

Dividing both sides of Equation 2.9 by SA and taking the limit as $\Delta t \to 0$ yields the differential form of the equation:

$$\frac{dH(t)}{dt} = -\frac{K}{SL_c}[H(t) - H_\infty]. \tag{2.10}$$

Equation 2.10 is called the *governing equation* of the system for the proposed conceptual model. It is a first-order ordinary differential equation that describes the rate of change of head in the basin with respect to time. A solution of Equation 2.10 will provide a function describing the change in head over time and, by extension, the temporal behavior of discharge from the basin through Equation 2.7.

The final item we need to complete our model is some type of condition specified on a model boundary. Roughly speaking, differential equations require one condition per order of each independent variable; since Equation 2.10 is a first-order differential equation in one independent variable (time), we require one condition for a complete solution. In this case, we will choose the condition that the head will be equal to some specified value at time $t = 0$:

$$H(t = 0) = H_0 \tag{2.11}$$

Because this condition is specified on the temporal boundary at time $t = 0$, it is called an *initial condition*.

2.4 Nondimensionalization

Although we have found a differential equation describing the rate of change of head with respect to time (Equation 2.10), much remains to be done in terms of understanding the characteristic behavior of the system. The first check that can be done on the model is to make certain the units on both sides of the equation match. This is the most basic test that any model must pass, and you should apply it to every model you develop. In this case, both sides of Equation 2.10 show units of $[L/T]$. Once dimensional consistency is established, the next step is to nondimensionalize the model.

There are numerous benefits of nondimensionalization (see Section 1.3.3), most of which will be illustrated in this and the following chapters. The process of nondimensionalization is not difficult, but a certain amount of practice is required to learn how to make appropriate choices when nondimensionalizing. A stepwise heuristic approach to nondimensionalizing equations is presented in Appendix A; here, we demonstrate the process by example.

The initial goal of nondimensionalization is to reduce the variables and the terms of the equation to order 10^0; that is, if possible we would like the variables to range from 0 to 1. We achieve this by normalizing the variables on appropriate values. Whenever possible, we use the values of specified boundary or initial conditions, along with our understanding of the extreme values the system may take, to normalize the variables. In this case, we can see from Equation 2.11 that one extreme value of head is likely to be H_0, the initial (and undoubtedly the greatest) value of head in the system. Since it is assumed there will be no precipitation after time $t = 0$, we can assume that head (and discharge) will decline over time, ultimately falling to H_∞ (see Figure 2.1). These considerations suggest the following normalization scheme for the dependent variable (head):

$$\theta = \frac{H(t) - H_\infty}{H_0 - H_\infty}. \tag{2.12}$$

Equation 2.12 fulfills our expectation for a nondimensional variable. The dimensionless head (θ) will vary from 1 (when $H = H_0$) to 0 (when $H = H_\infty$); all other values of θ should fall between these limits. In order to make a change of variables in Equation 2.10, however, we need to solve Equation 2.12 for H and calculate the derivative of H:

$$H = (H_0 - H_\infty)\theta + H_\infty, \tag{2.13}$$

$$dH = (H_0 - H_\infty)d\theta. \tag{2.14}$$

We now turn our attention to the independent variable, time. For this variable, it is not immediately apparent what quantities we should use for normalization. Time already begins at $t = 0$, but it runs to arbitrarily large values (conceivably, we may even be interested in the value of θ in the limit as $t \to \infty$). Since there is no characteristic quantity readily found in the auxiliary conditions (in this case, the initial condition), we follow the second rule of Appendix A and introduce an arbitrary characteristic time parameter t_c, with the intention of defining t_c in the near future. We will call the new dimensionless time variable τ,

$$\tau = \frac{t}{t_c}. \tag{2.15}$$

Solving Equation 2.15 for t and taking the derivative gives

$$t = t_c\tau, \tag{2.16}$$

$$dt = t_c d\tau. \tag{2.17}$$

Equations 2.13, 2.14, and 2.17 can now be substituted into Equation 2.10:

$$\frac{(H_0 - H_\infty)}{t_c}\frac{d\theta}{d\tau} = -\frac{K}{SL_c}(H_0 - H_\infty)\theta. \tag{2.18}$$

Dividing both sides by $(H_0 - H_\infty)$ and multiplying both sides by t_c gives:

$$\frac{d\theta(\tau)}{d\tau} = -\frac{Kt_c}{SL_c}\theta(\tau). \tag{2.19}$$

We have now reached a point where we can consider assigning a value to the characteristic time constant t_c. Inspection of Equation 2.19 reveals the units of K/SL_c to be $[1/T]$; since the units of characteristic time must be $[T]$, it appears we can achieve a great simplification of the governing equation if we allow $t_c = SL_c/K$. In fact, if we do this, Equation 2.19 reduces to

$$\frac{d\theta}{d\tau} = -\theta. \tag{2.20}$$

Applying the appropriate substitutions to the initial condition (Equation 2.11) yields the dimensionless initial condition:

$$\theta(\tau = 0) = 1. \tag{2.21}$$

Equations 2.20 and 2.21 are the final nondimensional forms of the governing equation and the initial condition.

The vastly simplified forms of Equations 2.20 and 2.21 (in contrast to the dimensional forms given in Equations 2.10 and 2.11) are the source of some confusion for analysts that are new to nondimensionalization. It is important to point out that *no information has been lost* in nondimensionalizing the governing equation. Careful comparison of the dimensional and dimensionless forms of the equations should convince the reader that the *forms* of the equations have not changed—nondimensionalization has only changed the scaling of the dependent and independent variables. Furthermore, the dimensional results can be recovered from the dimensionless results at any time by applying a back-transformation of the variables using Equations 2.12 and 2.15. The argument can in fact be made that we have gained information from the process of

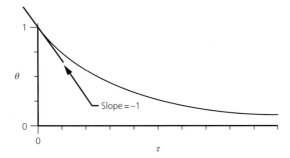

Figure 2.2 Schematic representation of the function $\theta(\tau)$.

nondimensionalization, because we have learned several important facts about the relationship among the dimensional parameters in the governing equation. For example, it should be clear from inspection of the characteristic time constant that changing the hydraulic conductivity by a factor of, for example, 2 cannot be distinguished from a change in the storativity or characteristic length scale of a factor of 1/2, since these operations all result in doubling the characteristic time. As a result, *it is not possible to distinguish uniquely the values of the individual parameters from the data alone.* If none of the three quantities K, S, or L_c are known *a priori*, the best that can be done is to uniquely identify the characteristic time of the system.[3]

We can gain other information from inspection of Equation 2.20 as well. We know from the initial condition (Equation 2.21) that $\theta = 1$ at $t = 0$, and substituting this information into Equation 2.20 shows us that the slope (i.e., the first derivative) of θ at $t = 0$ will be -1. Thus, θ will be a decreasing function: as θ decreases, the slope of θ will also decrease, asymptotically approaching 0 as $t \to \infty$. This curve is sketched schematically in Figure 2.2. Our qualitative sketch is in good agreement with the way we conceptualize discharge (and head) from the basin. We should keep our understanding of the form of the solution in mind as we seek a quantitative solution to Equation 2.20 in Section 2.5; if the form of our quantitative solution accords with our qualitative understanding of how the function behaves, we gain confidence in our solution. On the other hand, if our quantitative solution is at odds with our qualitative understanding, we must determine if our qualitative understanding is incorrect (and why that would be so), or if we have made an error in our solution.

2.5 Solving for θ

Although the governing equation (2.20) quantifies the rate of change in (nondimensional) head with respect to time, the equation can perhaps best be thought of as a kind of riddle: what function, when substituted for θ in Equation 2.20, will render the expression true? This function θ is the quantitative expression we seek. In some cases, the solution to a particular differential equation is relatively easy to find; in other cases, it may be very difficult, and there are any number of differential equations for which no known analytical solution exists. Fortunately, Equation 2.20 belongs to the category of differential equations that can be solved by a process of direct integration, and the solution is well within the capabilities of any student with a course in integral calculus.

To solve Equation 2.20, we divide both sides of the equation by θ so that all the θs are on the same side:

$$\frac{1}{\theta}\frac{d\theta}{d\tau} = -1. \tag{2.22}$$

We can now integrate both sides of the equation with respect to the independent variable τ:

$$\int \frac{1}{\theta}\frac{d\theta}{d\tau}d\tau = -\int d\tau. \tag{2.23}$$

We can change the variable of integration on the left-hand side, leaving

$$\int \frac{d\theta}{\theta} = -\int d\tau. \tag{2.24}$$

Performing the integrals yields:

$$\ln\theta = -\tau + C. \tag{2.25}$$

In Equation 2.25, we have lumped the constants of integration from both sides together into one constant (C). To get an explicit solution for θ as a function of τ, we can raise both sides of Equation 2.25 to a power of e:

$$e^{\ln\theta} = \theta = e^{-\tau+C} = e^{-\tau}e^{C} = Ce^{-\tau}. \tag{2.26}$$

All that remains is to find the constant of integration. To do this, we apply the initial condition (Equation 2.21). When $\tau = 0$, $\theta = 1$, and therefore it must be that $C = 1$. The final form of our model can now be written:

$$\theta(\tau) = e^{-\tau}. \tag{2.27}$$

A plot of Equation 2.27 is shown in Figure 2.3.

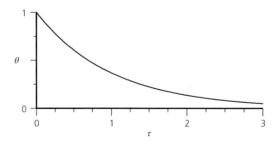

Figure 2.3 A plot of the function $\theta(\tau)$ as defined by Equation 2.27.

2.6 Calibrating the model to the data

Before we proceed further, we must first examine the model to see if the results make sense. We can check this in a number of ways; first, we should be sure the model meets the initial condition given in Equation 2.21. We do this by setting $\tau = 0$ in Equation 2.27, and looking to see if $\theta = 1$ (as stated in the initial condition), which it does. We also can examine the behavior of θ in the limit as $\tau \to \infty$. In this case, we find that $\theta \to 0$ as $\tau \to \infty$, which accords with our physical intuition about the problem of interest—namely, that the discharge from the basin should eventually go to zero if there is no additional precipitation. Finally, if we compare Figures 2.2 and 2.3, we can see that our qualitative solution (Figure 2.2) and our quantitative solution (Figure 2.3) are in excellent agreement. Thus, we have reasonable cause to believe our quantitative model is a good representation of our conceptual understanding of the system.

It can, of course, be argued that Equation 2.27 is an expression for (dimensionless) head as a function of time, rather than discharge, which is the quantity we are seeking. However, if we nondimensionalize the linearized Darcy's law (Equation 2.7) by normalizing on the initial discharge from the basin (i.e., the discharge at $\tau = 0$), we find that

$$\frac{Q}{Q_0} = \frac{-\frac{K}{SL_c}[H - H_\infty]}{-\frac{K}{SL_c}[H_0 - H_\infty]}, \tag{2.28}$$

$$\frac{Q}{Q_0} = \frac{H - H_\infty}{H_0 - H_\infty} = \theta, \tag{2.29}$$

where Q_0 is the discharge at time $t = 0$.

From Equation 2.29, it is clear that θ represents not only the nondimensional head but also the nondimensional discharge. This type of equivalence is not uncommon in nondimensional equations, and when it arises, it is extremely useful, because it means we have a choice of data to which we can apply the model. In this case, we do not have head data to calibrate the model, but we do have discharge data. As a result, we can use the historic discharge record to find the characteristic time constant t_c, which we can then apply to the model to make predictions about future discharge from the basin.

How can we find the characteristic time constant from historic data? There are at least two relatively straightforward ways to do so. Both methods depend on the physical meaning of the characteristic time: when the dimensional time is equal to the characteristic time, $\tau = 1$, the discharge will have dropped by a factor of $1/e$, and θ will therefore be equal to $\theta(\tau = 1) = 1/e \approx 0.37$. Thus, if we can determine the time it takes for the discharge from the basin to decrease to 37% of the initial value, we will know the characteristic time.

2.6.1 Semilog plots

The first method for determining the characteristic time is to graph the data on a semilog plot, with dimensional time [T] on the horizontal axis and the normalized discharge, Q/Q_0, on the vertical axis. The vertical axis should be the one plotted as a logarithm. It can be seen from Equation 2.25 that an exponential decay process will plot as a straight line on a semilog plot (with $\ln \theta$ as the ordinate). Once a straight line is drawn through the data, it is only necessary to find the value $Q/Q_0 = 0.37$ on the vertical axis, and note the corresponding dimensional time on the horizontal axis, as is shown in Figure 2.4 for a synthetic data set having a characteristic time of 55 days.

The method for finding the characteristic time described in the preceding paragraph is familiar to most groundwater hydrologists because of its use in interpreting slug test data (the perturbation to the potentiometric surface induced by the slug returns to static water level through a process of exponential decay). It is also possible to find the characteristic time from the slope of the line on the log-linear plot; however, it is important to note that most software packages plot the logarithmic axes of plots as base 10, while the logarithm of θ in Equation 2.25 is base-e (a natural logarithm). As a result, it is necessary to divide the slope of the plotted line (e.g., from Figure 2.4) by a factor of 2.3026. For example, reading the approximate points $(0, 1.0)$ and $(126.5, 0.1)$ from Figure 2.4, we can find the slope of the line:

$$\frac{\log_{10} \theta_1 - \log_{10} \theta_2}{t_1 - t_2} = \frac{0 - (-1)}{0 - 126.5} = -0.0079. \tag{2.30}$$

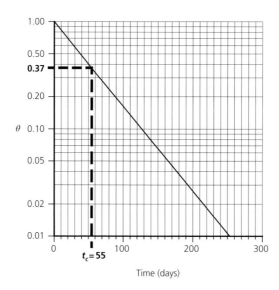

Figure 2.4 Illustration of the determination of characteristic time by the semilog method. The characteristic time of 55 days is identified as the point at which the dimensionless discharge $(\theta = Q/Q_0)$ has decreased to ≈ 0.37 of its initial value.

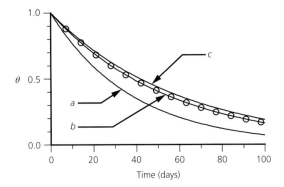

Figure 2.5 Determination of characteristic time by trial and error. The curve labeled *a* has a characteristic time of 38 days; the curve labeled *b* has a characteristic time of 55 days (the best fit curve); and the *c* curve has a characteristic time of 60 days. The data, plotted once per 7 days, are shown as circles.

Taking the reciprocal of the absolute value (1/0.0079) and dividing by 2.3026 yields a characteristic time of 54.938, or ≈55 days.

2.6.2 Curve matching

The second method for determining the characteristic time of the system is slightly more complex, but, because it uses trial and error, it may be more satisfying to many modelers. The process is readily accomplished with one of the many spreadsheet programs that are available as prepackaged software for personal computers. The modeler first plots the data on a set of axes with time as the abscissa (i.e., on the horizontal axis) and normalized discharge (i.e., with all discharges divided by the initial discharge) as the ordinate (the vertical axis). The exponential decay curve:

$$\theta = e^{-t/t_c} \tag{2.31}$$

is plotted on the same axes as the data. The quantity t_c is then varied until the modeled exponential decay line coincides with the plotted data; this can either be achieved by eye, or by some method of minimizing the residuals. Once a satisfactory match has been made, the characteristic time can be read directly from the spreadsheet (i.e., the fitted value of t_c). This method is shown in Figure 2.5, where a number of sample curves are plotted along with the data to illustrate several possible characteristic time curves.

2.7 Extending the model

A plot of dimensionless discharge for the Bandurraga Basin is shown in Figure 2.6, along with two exponential decay curves that were calculated in an attempt to fit the data using the curve matching method. An examination of the figure

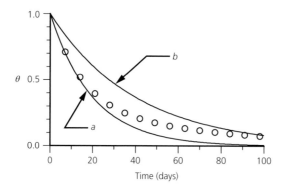

Figure 2.6 Attempting to determine the characteristic time by trial and error for a synthetic data set representing the Bandurraga Basin, Idaho. The curve labeled *a* has a characteristic time of 20 days; the curve labled *b* has a characteristic time of 40 days. Neither curve fits the data, which are shown as circles (data shown once per 7 days).

reveals that neither curve fits the data well. In fact, numerous trial-and-error attempts to fit the data demonstrate that no exponential decay curve will fit the data plotted in the figure, regardless of the characteristic time tested. Why doesn't our model fit the data, and is it possible to somehow modify or extend the model to better represent the data?[4]

If we were to examine the data given in the list of known quantities (Section 2.3.3), we may note the item "information on soil types and thicknesses." It turns out that the lower reaches of our basin contain an extensive old-growth cedar grove. Such a grove is characterized by thick, well-developed soils with high organic content; consequently, these soils have high water storage and retention properties. Considering the likely high storativity and moderate hydraulic conductivity, we can expect the characteristic time of this part of the basin to be relatively long. The remainder of the basin is characterized by thin or absent soil cover, and the majority of the water is contained within fracture networks in the granitic bedrock. This fractured bedrock part of the basin may have high permeability, but the storativity will be low. As a result, we can expect the fractured rock portion of the basin to have a short (in comparison to the old-growth cedar grove) characteristic time.

The reason that our model cannot adequately represent discharge from the basin should now be clear: the basin effectively comprises two sub-domains with different characteristic times. A one-domain model such as that given in Equation 2.27 cannot be expected to correctly predict the behavior of such a system. But can we modify our model to account for this? In fact, we can use the principle of superposition to extend our model to cover the two-domain case. The principle of superposition states that, for a linear system, the response of a system to two (or more) stimuli is equal to the sum of the responses of the system to each individual stimulus.

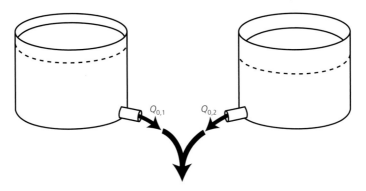

Figure 2.7 Conceptual model for the two-domain representation of the Bandurraga Basin.

The revised conceptual model of the Bandurraga Basin is shown in Figure 2.7. We now have two domains that yield water to the main stem; the discharge that is measured is the sum of the individual discharges from the two domains. Dimensionally, we can quantify this as:

$$Q = Q_{0,1}e^{-t/t_{c,1}} + Q_{0,2}e^{-t/t_{c,2}}, \tag{2.32}$$

where the subscripts "0, 1" and "0, 2" imply the initial discharge from the first and second domains, respectively, and the subscripts "c, 1" and "c, 2" similarly denote the characteristic times of the first and second domains. If we nondimensionalize by dividing both sides of the equation by the sum of the initial discharge from the two domains, we have:

$$\theta = \frac{Q}{Q_{0,1} + Q_{0,2}} = \frac{Q_{0,1}}{Q_{0,1} + Q_{0,2}}e^{-t/t_{c,1}} + \frac{Q_{0,2}}{Q_{0,1} + Q_{0,2}}e^{-t/t_{c,2}}. \tag{2.33}$$

We can write this in a more compact form as:

$$\theta = \phi_1 e^{-t/t_{c,1}} + \phi_2 e^{-t/t_{c,2}}, \tag{2.34}$$

where

$$\phi_1 = \frac{Q_{0,1}}{Q_{0,1} + Q_{0,2}}, \tag{2.35}$$

$$\phi_2 = \frac{Q_{0,2}}{Q_{0,1} + Q_{0,2}}. \tag{2.36}$$

It should be clear, however, that

$$\phi_1 + \phi_2 = 1. \tag{2.37}$$

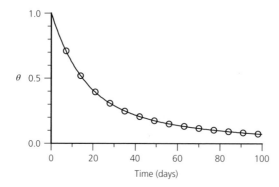

Figure 2.8 A plot of the same data shown in Figure 2.6, fit with a two-domain exponential decay model. The model as fit uses the parameters: $\phi = 0.7$, $t_{c,1} = 15$ days, and $t_{c,2} = 70$ days.

Furthermore, it is inconvenient to have two characteristic times in the same equation. We can fix this by introducing a parameter α, which we will define as the ratio of characteristic times of the two domains:

$$\alpha = \frac{t_{c,1}}{t_{c,2}}. \tag{2.38}$$

Substituting Equation 2.37 into Equation 2.34 and multiplying the exponent of the second term by $t_{c,1}/t_{c,1} = 1$, we can rewrite the equation as:

$$\theta = \phi_1 e^{-t/t_{c,1}} + (1 - \phi_1)e^{\frac{t_{c,1}}{t_{c,2}} \frac{-t}{t_{c,1}}}, \tag{2.39}$$

$$\theta = \phi e^{-\tau} + (1 - \phi)e^{-\alpha\tau}. \tag{2.40}$$

In Equation 2.40, we have dropped the subscript on ϕ, and both terms of the equation use the same τ (normalized on the characteristic time of the fractured rock domain). The difference in characteristic time between the two domains is accounted for by the use of the α parameter. This two-domain model is somewhat more difficult to calibrate than the one-domain model because of the additional parameters, but the calibration can still be accomplished using the same curve fitting method outlined for the one-domain model in Section 2.6.2. Note that, although the nondimensional model (Equation 2.40) is characterized by two parameters, in order to fit the model, three dimensional parameters must be estimated: the fraction of discharge resulting from one of the two domains (the fraction of water from the other domain can be found from Equation 2.37) and the characteristic times of the two domains. A two-domain model fit to the data plotted in Figure 2.6 is shown in Figure 2.8.

2.8 A numerical solution for exponential decay

Before moving on from the exponential decay problem, it will be worthwhile to spend a few minutes developing a method for finding an approximate numerical solution to the problem. Although, in this case, we were able to find an exact analytical solution (Equation 2.27) to Equation 2.20, it often happens that an analytical solution to a problem of interest may be difficult or impossible to find. By taking the time to develop a numerical method of approximating the solution to a simple problem, we will have the tools to solve these more difficult problems when they arise.

Recall the derivative of a function $f(t)$ is defined as:

$$\frac{df(t)}{dt} = \lim_{\Delta t \to 0} \frac{f(t + \Delta t) - f(t)}{\Delta t}. \tag{2.41}$$

We can use Equation 2.41 to find an approximation to the derivative of an arbitrary function. Instead of taking $\lim_{\Delta t \to 0}$, we will stipulate that Δt remain small, but finite. For sufficiently small Δt, we have

$$\frac{df(t)}{dt} \approx \frac{f(t + \Delta t) - f(t)}{\Delta t}, \tag{2.42}$$

or, to put Equation 2.42 in terms of our nondimensional variables:

$$\frac{d\theta}{d\tau} \approx \frac{\theta^{i+1} - \theta^i}{\Delta \tau}. \tag{2.43}$$

In Equation 2.43, the superscript on θ is the "timestep index" (here designated as i) that denotes the current timestep, where $i = 0$ is the initial time, $i = 1$ is time $\tau = \Delta \tau$, and so on. The formula for calculating the time represented by any arbitrary i for a constant timestep is:

$$\tau = i\Delta \tau. \tag{2.44}$$

Equation 2.43 is known as a "finite difference" approximation for the first derivative. We can substitute the finite difference approximation of the derivative into our equation for the model of the Bandurraga Basin (Equation 2.20) to obtain an approximate numerical representation of the problem:

$$\frac{\theta^{i+1} - \theta^i}{\Delta \tau} = -\theta^{i+1}. \tag{2.45}$$

Note that we have chosen to use the value of θ at the future timestep for the right-hand side of Equation 2.45 (we could also have used the value of θ at this timestep). We can solve Equation 2.45 for θ^{i+1} in terms of θ^i:

$$\theta^{i+1} = \frac{\theta^i}{1 + \Delta \tau}. \tag{2.46}$$

We know that $\theta = 1$ at $\tau = 0$, which is equivalent to $i = 0$, from the initial condition. As a result, we can use Equation 2.46 to step through any number of timesteps to find the value of θ at any desired τ.

Note there is an implicit assumption in the finite difference formulation (Equations 2.42 and 2.43): we are assuming the derivative can be approximated as a linear function. If we could take the limit as Δt goes to zero the derivative would be exact, but we can't do this on a computer. In effect, we are using the first term of a Taylor series to represent our function; we therefore say our finite difference scheme is accurate to the "first order." (If we used more terms in the Taylor series, we would say the scheme was "second order," "third order," etc., if we used two terms, three terms, etc.) Because we are using a first-order scheme, we must keep our timesteps small for the linear approximation to be a good one.

But what is a "small" timestep in this context? That depends on the function we wish to represent. If the first derivative of the function is changing rapidly (i.e., if the second derivative of the function is large), we may need very small timesteps indeed. If the first derivative of the function is changing very slowly (if the second derivative is small), we may be able to use quite large timesteps. In fact, it is often possible to vary the timestep size over the domain of interest. Commonly, we use very small timesteps near the initial condition when change is rapid, and gradually increase the timestep size as the system approaches steady state (unless other things are going on that disturb the approach to equilibrium). Most finite difference modeling software has the option to increase timestep size according to some criterion (e.g., the timestep may be increased according to some predetermined schedule, as long as the absolute or relative error remains below some user-determined amount).

The last thing we should note in this section is that the finite difference operator defined by Equation 2.43 is completely general. That is, its use is not restricted to nondimensional variables, or even to the use of time (or dimensionless time) as the independent variable. Any dependent and independent variables can be substituted for θ and τ in Equation 2.43. This finite difference approximation can be used to represent any first-order derivative, as long as the function is "sufficiently" smooth and the timestep is "sufficiently" small. Commonly, we differentiate between spatial and temporal derivatives by putting the index in the superior position for temporal derivatives and in the inferior position for spatial derivatives, as will be seen in examples in the following chapters.

2.9 Conclusions

In this section, I presented a basic approach to the development of a mathematical model of a physical system. To begin with, the modeler should define exactly the goal(s) of the model (the "FIND" statement). Subsequently, the modeler should list the known information about the system and the assumptions that will have to be made to develop a tractable conceptual model. Once these steps are done, the modeler clearly states the approach to model development and implements the solution in a rational, stepwise manner. This approach to model development was illustrated by examining the problem of predicting discharge from a hypothetical alpine basin in Idaho. A mass balance approach was used to represent the basin as a lumped system; the result of this analysis was an exponential decay (or baseflow recession) model. A scheme for nondimensionalizing the governing equation was presented, and the method of direct integration was used to solve the nondimensional governing equation. Two methods for calibrating the model were discussed, and a method (superposition) that allowed the extension of the model to a more complex situation was presented. Finally, we examined a simple "finite difference" scheme for arriving at an approximate numerical solution to the governing equation.

Although the model presented is very simple, it is quite general, and may apply to a great many problems both in hydrology and in other branches of science and engineering. As was shown in the two-domain case, the model is readily extended to cover wider ranging and more complex situations; for example, Baedke and Krothe (2001) describe a three-domain baseflow recession model of discharge in karst terrain.

Beyond its direct applications, the examples presented in this chapter demonstrate many of the basic principles of model development, and will serve as a template for the construction of more complex models in later chapters. In addition, the finite difference scheme introduced in this chapter forms the basis for many of the groundwater models (and other types of models) currently in use in the scientific and engineering communities. Although other types of numerical models are also popular (e.g., finite element models and boundary element models), a solid understanding of finite differences will provide a good background for the application of most numerical models, and for advanced study of other numerical techniques. Furthermore, the interested student will find it relatively easy to write computer code for one's own finite difference programs. The conceptual, analytical, and numerical principles outlined in this chapter will be pursued and expanded on in the following sections.

2.10 Problems

1. Use the one-domain exponential decay model to determine when the water manager for the Bandurraga Basin should begin storing water in order to have a full reservoir on August 1. The discharge on May 1 is 33,000 m^3/day, and the characteristic time of the basin is 55 days. The reservoir has a capacity of 45,000 m^3. You may neglect the amount of water needed by the hydroelectric generator each day to maintain power at the visitor's center.

2. Repeat problem 2.10, allowing for the 1000 m^3/day needed by the hydroelectric generator. HINT: you may want to read over the material in Appendix B.

3. Use the finite difference model (Equation 2.46) to calculate the curve $\theta = e^{-\tau}$ for values of τ from 0 to 5. Compare your calculated results to the actual values. Try using three different timestep sizes; for example, you may want to try 0.5, 0.05, and 0.01 (you may use other values if you like). How does the numerical approximation compare with the analytical solution? How small do you have to make your timesteps to get a good approximation? Show your results graphically and make a table comparing *selected* values (do not just print out the answers for all the timesteps).

4. Can you think of a way to use the finite difference formula (Equation 2.46) to derive a single equation that approximates the value of the exponential decay curve at any arbitrary timestep? HINT: Apply the formula recursively.

5. The basic exponential decay model (Equation 2.20) was extended by Fairley (2003) to model Inka water resource management methods in pre-Columbian Peru. The underlying idea is to represent the basin with an exponential decay model that discharges into a storage area that is also governed by an exponential decay model (although having different parameters from the larger basin). The final dimensionless governing equation[5] is given by

$$\theta = \theta_0 e^{-\tau} + \frac{\beta}{1 - \alpha} \left(e^{-\alpha\tau} - e^{-\tau} \right), \tag{2.47}$$

where θ_0 is the initial amount of water in the constructed storage system. See if you can rederive the Fairley (2003) model for "geological water storage." What do the dimensionless parameters α and β represent?

Notes

1 Although equivalent formulations of the same principle can be found in writings as far back as Pythagoras.

2 Note that, in canceling these terms, we are implicitly assuming constant density. This assumption should therefore be added to our list.

3 There may be other models or methods of analysis that could be used to find the parameter values (e.g., statistical analyses and drilling wells), but fitting the data to the model given in Equation 2.10 cannot uniquely identify the values of K, S, or L_c.

4 The idea for this extension of the basin discharge model was suggested to me by D. Hopster, an MSc student in my modeling class at the University of Idaho.

5 Note that there is a typographical error in the equation as it appears in the original article.

References

Badius, J. and Trechsel, J. (1495) *Quaestiones et decisiones in quattuor libros Sententiarum Petri Lombardi: Centilogium theologicum*, Johannes Trechsel, Lyon.

Baedke, S.J. and Krothe, N.C. (2001) Derivation of effective hydraulic parameters of a karst aquifer from discharge hydrograph analysis. *Water Resources Research*, **37** (1), 13–19.

Fairley, J.P. (2003) Geologic water storage in pre-Columbian Peru. *Latin American Antiquity*, **14** (2), 193–206.

CHAPTER 3
A model of water quality

Chapter summary

In this chapter, we use the principles introduced in Chapter 2 to develop a model for the water quality in an artificial lake, located in southern Nevada. The model, which is a classic "salt tank" problem from most introductory differential equations classes, is useful for examining the consequences of some basic water quality management decisions. In addition, the method demonstrated forms one end member of the spectrum of techniques used, for example, to estimate groundwater age dates from isotopic samples and similar problems.

3.1 Oases in the desert

During the late 1980s and 1990s, the city of Las Vegas, Nevada, experienced an enormous growth in population, and a commensurate explosion of home building. In addition to building of apartment houses, much of the construction was centered on tract housing for middle- and upper-middle-class prospective homeowners. In a (largely successful) bid to command top dollar for high-end homes, some builders began constructing artificial lakes as centerpieces for their housing developments. Projects such as "The Lakes" on the west side of Las Vegas and the "Lake Las Vegas" community in Henderson, Nevada, about 17 miles east of the Las Vegas strip, used the aesthetic appeal of water in the desert to attract well-to-do home buyers to their artificial oases. Other, similar lake/real estate developments can be found throughout the arid southwestern United States, in states such as California, Nevada, and Arizona.

Naturally, the developers of these real estate projects want to minimize the costs of managing water quality in the artificial lakes, as would the Homeowner's Associations (HOAs) that are responsible for maintaining the aesthetic appeal of the lakes after the houses have all been sold. This economic motivation for water quality management in an artifical lake can be translated into the following imperative: to minimize the costs of maintaining the lake (i.e., to use the minimum quantity of makeup water necessary) while meeting the community water quality goals. In order to balance these competing interests (minimum

Models and Modeling: An Introduction for Earth and Environmental Scientists, First Edition. Jerry P. Fairley.
© 2017 John Wiley & Sons, Ltd. Published 2017 by John Wiley & Sons, Ltd.
Companion website: www.wiley.com/go/Fairley/Models

water use and maintaining water quality), it is necessary to have a framework within which to make decisions about water management. The development of such a framework is the subject of this chapter.

3.2 Understanding the problem

There are, of course, many ways in which the water quality goals of a community may be defined. For example, a community may want the lake to support a particular type of game fish, or be attractive for swimmers. More quantitatively, a community may set goals for water temperature, clarity, dissolved oxygen content, or many other measurable parameters. Most likely, several of these indicators of water quality would be tracked to maintain the lake in an acceptable condition. One important indicator of water quality at real estate developments is the concentration of total dissolved solids (TDS). TDS became of special interest to water managers of artificial lakes in arid regions around 2010, because some golf courses associated with the real estate developments refused to continue to purchase water from the HOAs. These golf courses, which had previously been irrigated with lake water, attempted to negotiate their own contracts for (raw) municipal water based on the claim that high TDS concentrations made lake water unsuitable for irrigating their properties. Whether or not these claims are defensible, it is clear that it is in everyone's interests to maintain a high standard of water quality for aesthetic and practical (economic) reasons. In order to balance economic and water quality factors, we require a model that will allow us to test various management strategies on TDS levels in these human-made lakes.

3.3 Model development

At this point, we probably have enough information to begin to formulate the model. We will use the method illustrated in Chapter 2 to develop the current model; that is, we will make a brief, concise statement of that which we want to find, make lists of our knowns and assumptions, and give a description of our approach. Once we have completed all these tasks, we can carry out our approach and apply the resulting model to our water quality management problem.

FIND: An expression that gives concentrations of TDS (in mg/l or similar units) as a function of time for an artificial lake in southern Nevada. The model should allow us to test various management options (i.e., adding more or less makeup water) on water quality.

KNOWN:

- Volume and surface area of the lake
- Incoming water quality (i.e., TDS concentration of incoming water)
- Potential (pan) evaporation rate

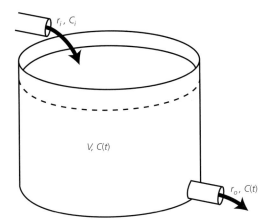

Figure 3.1 Conceptual sketch for a salt tank model of water quality in an artificial lake.

ASSUME:

- Complete mixing in the lake
- No dissolution of minerals from the lake bottom or precipitation from lake water
- Constant input water quality (including no influx of groundwater–although discharge to groundwater is OK)
- Constant lake volume (including "instantaneous" filling of the lake)
- Initial TDS concentration equal to input water TDS

APPROACH: Conceptualize the lake as a large tank, with incoming water at some (initial and constant) TDS and outgoing water equal in concentration to the volume-averaged concentration of dissolved solids in the lake. Use mass balance to formulate the governing equation and solve (this is a classic "salt tank" approach; see Figure 3.1).

3.3.1 Model formulation

The salt tank problem is well known from most ordinary differential equation courses. In essence, salt tank models assume some quantity is being carried into a reservoir at a known rate and concentration, and being carried away at the average concentration of the well-mixed reservoir. The problem is quite general, and can be applied not only to species concentrations (i.e., TDS) but also to heat energy (i.e., water temperature), isotope concentrations (e.g., for groundwater age determinations), and so on.

To formulate the model, we use the mass balance equation:

$$M_{\text{in}} = M_{\text{out}} + \Delta M. \tag{3.1}$$

The terms of the mass balance can be expanded as:

$$M_{in} = r_i C_i \Delta t, \tag{3.2}$$

$$M_{out} = r_o C(t) \Delta t, \tag{3.3}$$

$$\Delta M = \Delta[VC(t)]. \tag{3.4}$$

In these equations, r_i and r_o represent the rates of water coming into and out of the reservoir, respectively $[L^3/T]$; these rates only include water that is either bringing in or taking away dissolved solids. In particular, the water lost from the lake by evaporation is not included in the r_o term. C is the concentration of TDS $[M/L^3]$; the subscript i indicates the initial concentration of TDS in the lake (equal to the concentration of TDS in the incoming water), and the concentration term with explicitly displayed time dependence ($C(t)$) indicates the concentration in the well-mixed lake (i.e., the average concentration of TDS in the lake). V is the volume of the lake $[L^3]$, and t is time $[T]$. A check of the units of the three equations makes it clear the units of each are those of mass $[M]$, as should be the case.

Substituting Equations 3.2, 3.3, and 3.4 into the mass balance equation gives:

$$r_i C_i \Delta t = r_o C(t) \Delta t + \Delta[VC(t)]. \tag{3.5}$$

Because the volume of the lake is assumed constant (see the list of assumptions), we can pull V out from under the Δ in Equation 3.5. Dividing both sides of the equation by V and Δt yields:

$$\frac{\Delta C(t)}{\Delta t} = \frac{r_i C_i - r_o C(t)}{V}. \tag{3.6}$$

Taking the $\lim_{\Delta t \to 0}$, we get

$$\frac{dC(t)}{dt} = \frac{r_i C_i - r_o C(t)}{V}, \tag{3.7}$$

with an initial condition,

$$C(t = 0) = C_i. \tag{3.8}$$

3.3.2 Nondimensionalization

Equations 3.7 and 3.8 are the dimensional forms of the governing equation and initial condition. We would now like to nondimensionalize the governing equation; however, nondimensionalizing the dependent variable (concentration) here is a little different than in the Bandurraga Basin problem because, although we would still like the dimensionless concentration to be of order 10^0, we would like the initial concentration to equal 1. Higher concentrations would then be greater than 1, and lower concentrations would be less than 1. To accomplish this, we choose

$$\theta = \frac{C(t)}{C_i}. \tag{3.9}$$

Solving for $C(t)$ and taking the derivative gives

$$C(t) = C_i \theta, \tag{3.10}$$

$$dC(t) = C_i d\theta. \tag{3.11}$$

In the Bandurraga Basin problem, we had no clear characteristic time quantity on which to normalize time, and the same is true for this problem. As a result, we insert an arbitrary characteristic time t_c. Defining τ, solving for t and its derivative yields

$$\tau = \frac{t}{t_c}, \tag{3.12}$$

$$t = t_c \tau, \tag{3.13}$$

$$dt = t_c d\tau. \tag{3.14}$$

Substituting Equations 3.10, 3.11, 3.13, and 3.14 into Equations 3.7:

$$\frac{C_i}{t_c} \frac{d\theta}{d\tau} = \frac{r_i}{V} C_i - \frac{r_o}{V} C_i \theta, \tag{3.15}$$

$$\frac{d\theta}{d\tau} = \frac{r_i t_c}{V} - \frac{r_o t_c}{V} \theta. \tag{3.16}$$

At this point, it should be clear that the characteristic time can be set equal to either V/r_i or V/r_o, both of which have units of [T]. However, the question is, which should we pick: V/r_i or V/r_o?

This is a subtle point that deserves some thought. The volume of the lake, V, divided by the rate of replacement of water in the lake, is commonly referred to as the "residence time" of water in the lake. Which quantity, V/r_i or V/r_o, is more representative of the residence time of the lake? The quantity r_o is the outflow from the lake carrying a flux of dissolved solids; however, as was remarked earlier, this outflow does not include the evaporative flux out of the lake, which does not carry away dissolved solids. On the other hand, r_i is the total quantity of water that must be added to the lake to keep the volume constant (recall that constant lake volume was one of the assumptions of the model). As a result, r_i includes water to replace evaporation losses as well as losses to seepage and management outflows (water released to control the buildup of TDS in the lake). The ratio of the lake volume to the rate at which water is coming into the lake, V/r_i, is therefore likely to be a better indicator of the residence time. On the basis of this consideration, we will choose $t_c = V/r_i$. Substituting for t_c and nondimensionalizing the initial condition provides us with the nondimensional governing equation and initial condition:

$$\frac{d\theta}{d\tau} = 1 - \epsilon\theta, \tag{3.17}$$

$$\theta(\tau = 0) = 1. \tag{3.18}$$

The dimensionless parameter ϵ in Equation 3.17 is defined as $\epsilon = r_o/r_i$, and quantifies the rate at which water is carrying away dissolved solids (at the concentration of the lake) as a fraction of how fast new dissolved solids are being brought in (at the inflow concentration). It will be useful later to recognize that ϵ can also be written as

$$\epsilon = \frac{r_o}{r_i} = \frac{r_o}{r_o + r_e}, \tag{3.19}$$

where r_e is the rate at which water is leaving the lake by evaporation.

3.3.3 Solving the equation

Of course, in order to obtain the temporal behavior of θ (or concentration) explicitly, it is necessary to solve Equation 3.17. Can we solve this particular differential equation? As a matter of fact, we can, using the same technique we used to solve the exponential decay model (i.e., direct integration). First, we must divide both sides by the quantity $1 - \epsilon\theta$ to get all the terms containing θ on the same side of the equation. Integrating both sides with respect to τ gives us:

$$\int \frac{1}{1 - \epsilon\theta} \frac{d\theta}{d\tau} d\tau = \int d\tau, \tag{3.20}$$

$$\int \frac{d\theta}{1 - \epsilon\theta} = \int d\tau, \tag{3.21}$$

$$-\frac{1}{\epsilon} \ln(1 - \epsilon\theta) = \tau + A, \tag{3.22}$$

$$\ln(1 - \epsilon\theta) = -\epsilon\tau + A, \tag{3.23}$$

$$1 - \epsilon\theta = e^{-\epsilon\tau + A} = Ae^{-\epsilon\tau}, \tag{3.24}$$

$$\epsilon\theta = 1 - Ae^{-\epsilon\tau}, \tag{3.25}$$

$$\theta = \frac{1}{\epsilon} - Ae^{-\epsilon\tau}. \tag{3.26}$$

Here, we have absorbed the constants of integration from both sides into A. In addition, we have again taken advantage of the fact that $e^{-\epsilon\tau + A}$ can be written as $e^A e^{-\epsilon\tau}$, and recycled the constant A. It remains to find the constant A, which we can do by applying the initial condition (Equation 3.18) to find that $A = (\frac{1}{\epsilon} - 1)$. So the final solution is:

$$\theta = \frac{1}{\epsilon} - \left(\frac{1}{\epsilon} - 1\right)e^{-\epsilon\tau}, \tag{3.27}$$

$$\theta = \frac{1 - e^{-\epsilon\tau}}{\epsilon} + e^{-\epsilon\tau}, \tag{3.28}$$

$$\theta(\tau) = \frac{1 - e^{-\epsilon\tau} + \epsilon e^{-\epsilon\tau}}{\epsilon}. \tag{3.29}$$

A plot of the solution (Equation 3.29) is shown in Figure 3.2.

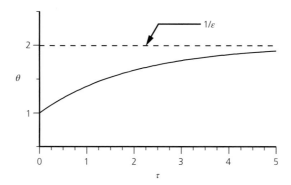

Figure 3.2 Plot of the solution (Equation 3.29) for $\epsilon = 0.5$.

3.4 Evaluating the model

Now that we have a solution, it is time to take stock and ask if our solution makes sense or not. How can we do that? First, we should check to make certain the initial condition is met (if not, it is likely we made an error in finding the value of the constant of integration). Substituting 0 in for τ shows that the initial condition is, in fact, met. Next, we should ask whether the solution performs according to our intuition in the limits—that is, what does the solution do as $\tau \to \infty$, or when ϵ takes on extreme values?

It is relatively easy to check the "final" value of θ, because

$$\lim_{\tau \to \infty} \theta = \lim_{\tau \to \infty} \frac{1 - e^{-\epsilon\tau} + \epsilon e^{-\epsilon\tau}}{\epsilon} = \frac{1}{\epsilon}. \tag{3.30}$$

I say "final" in quotes because we can see from Equation 3.29 that $\theta \to 1/\epsilon$ asymptotically as $\tau \to \infty$ (Figure 3.2). Although we don't know if $1/\epsilon$ is the correct value for steady-state water quality in the lake, it does seem reasonable that the long-term TDS would be some function of the ratio of input and outflow.

Next, we can look at the steady-state behavior of the model when ϵ takes on extreme values. The most extreme values of ϵ we can have in our model are 0 (when there is no outflow, and inflow is limited to replacement of evaporation losses) and 1 (when replacement is so rapid that evaporation losses are negligible in comparison). It is easy to see from Equation 3.30 that $\theta \to 1$ as $\tau \to \infty$ for the case of $\epsilon = 1$. Does this make sense? An analogous physical situation would be a short reach of a river: water is coming in at the upstream boundary of our reach at some TDS, passing through the reach with negligible loss to evaporation, and leaving the reach through the downstream boundary at about the same concentration of TDS. So the prediction by our model for the case of $\epsilon = 1$ is reasonable.

The case of $\epsilon = 0$ is more difficult, because Equation 3.30 goes to infinity as $\epsilon \to 0$. However, if we set $\epsilon = 0$ in Equation 3.17, we can re-solve by direct

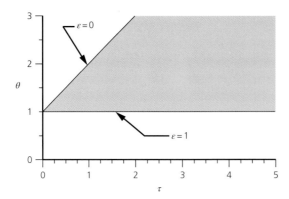

Figure 3.3 Plot of the solutions for the two extreme values of ϵ, where $\epsilon = 0$ or $\epsilon = 1$, as labeled. The shaded area indicates the possible solution space, assuming an initial and input concentration of $\theta = 1$.

integration and find the constant of integration from applying the boundary condition. If we do this, we find the solution:

$$\theta = \tau + 1. \tag{3.31}$$

It is clear from Equation 3.31 that the concentration goes linearly to infinity (in agreement with Equation 3.30). Only the losses to evaporation are being replaced; with no outflow to carry away dissolved solids, all the solids coming into the lake stay permanently. This is the situation that is found in a terminal lake (a lake that has no outlet), such as the Great Salt Lake in Utah, and our model also predicts this scenario perfectly. The limiting behavior for extreme values of ϵ is shown in Figure 3.3. The fact that the model apparently predicts correctly the behavior of the system in the limits of large and small ϵ does not guarantee that the model accurately reproduces the system behavior as a whole, but it does build confidence that the model output is reasonable.

3.5 Applying the model

Although the task of actually putting numbers in the model is left for the exercises (Section 3.7), a few words about the application of the model to the management of water quality are appropriate. How would we use this model to actually make decisions about water quality management strategies?

Perhaps the most important thing to note is that the model gives predictions of water quality at any arbitrary time, which includes a prediction of (dimensionless) TDS concentration at infinity. If we are concerned with maintaining a level of dissolved solids in the lake over a long timescale (say tens of years), we can use the prediction that $\lim_{\tau \to \infty} \theta = 1/\epsilon$ to plan our management strategy. Since we (presumably) know our target water quality, C_{target}, we can calculate the corresponding target θ using Equation 3.9. The reciprocal of θ_{target} is $1/\epsilon$, and from Equation 3.19 we have

$$\epsilon = \frac{r_o}{r_o + r_e}, \tag{3.32}$$

where r_e is the evaporation loss from the lake. Since we know two of the three variables in Equation 3.32 (ϵ and r_e, which is usually known at least approximately for a given region), we can readily calculate the rate of management withdrawal that is required to maintain the desired water quality indefinitely.

It may happen, however, that in some cases we are not interested in maintaining water quality indefinitely, but rather in reaching a particular target water quality at a target time. This problem is more awkward to handle nondimensionally, because we only know one of the three quantities necessary; that is, we know θ_{target}, but we do not know ϵ or τ_{target}. We don't know *a priori* the characteristic time of the system, because the characteristic time is based on the r_i or, ultimately, on r_o (through Equation 3.32), and this is the quantity we seek. In this case, the easiest thing to do is to re-dimensionalize the problem and solve for r_o. These issues are explored more deeply in the problems following this chapter.

3.6 Conclusions

In this chapter, we developed a model of water quality in an artificial lake in the arid southwestern United States. We used TDS as an indicator of water quality and formulated a mass balance model that can be applied to the management of dissolved solids when the ratio of incoming water to management outflows can be controlled. The model results show that, in nondimensional terms, TDS concentration asymptotically goes to $1/\epsilon$ as the lake approaches steady state. Furthermore, the extreme values of ϵ, which is the dimensionless ratio of TDS-carrying outflow from the lake to the total inflow, delineate the bounds on behavior. The two bounding cases, $\epsilon \to 0$ and $\epsilon \to 1$, are roughly equivalent to a terminal lake (a hypersaline body of water) and a continually flushed body of water (e.g., a stream reach) of constant quality, respectively.

Apart from any usefulness the water quality model may have for managing TDS in artificial lakes in the desert southwest, variations on the salt tank model are commonly used in the chemical engineering field, where they are known as "well-stirred reactor" models. The model is also important for the role it plays in understanding groundwater residence times in aquifers. This is because age dates from ^{14}C, tritium, and other isotope-based age dating methods cannot be taken uncritically, but must be evaluated against some model of mixing in the aquifer. Two models form the end-members in this evaluation: a "plug flow" model, in which water moves through the aquifer without any mixing, and a complete mixing model that is in essence the model developed in this chapter. Because the two models represent extreme cases, the actual age of the water (i.e., the time since the last exposure to the atmosphere) should fall in between. The water quality model developed here can, therefore, be used to represent a number of important systems, and forms a useful tool in every hydrogeologist's toolbox.

3.7 Problems

The standard for drinking water quality is 1000 mg/l TDS for sole-source water (500 mg/l if there is an alternative available). Incoming water from Lake Mead used to manage the water quality of an artificial lake near Las Vegas, Nevada, is 629 mg/l TDS. The total volume of the lake is 10,600 acre-feet, and approximately 2500 acre-feet/year are lost to evaporation.

1. What rate of inflow must be maintained from the outset (i.e., $\tau = 0$) in order to keep the lake water within the sole-source drinking water quality standard forever?

2. Suppose you are the developer that built and sold the homes around the lake described before. You are contractually bound to maintain the water quality at sole-source drinking water standards for a term of 3 years. After 3 years, the HOA takes over management of the water quality. To maximize your profits, you would like to put the minimum possible water into the lake, while still meeting your obligation. At what rate would you add water to the lake? What is the final (steady-state) water quality at this rate?

3. Once the developer has moved on to greener pastures, the HOA takes over the management of water quality at the lake. The initial water quality at the time the HOA assumes responsibility is 1000 mg/l, the sole-source drinking water standard. One year after the hand-off, the HOA decides to return the lake to drinking water quality. Can the HOA achieve drinking water quality in the lake? What are the HOA's options for managing the water quality? What advice would you give them? How will their options impact the HOA finances? HINT: You cannot change ϵ in any given model, but you can reformulate the model with a new initial condition at a new initial time $\tau = 0$; you will need to recalculate the constant of integration for the new initial condition.

4. Re-derive the model to use concentrations in mg/kg rather than mg/l. Does this new formulation of the model significantly change the model structure or predictions?

5. Apply the finite difference scheme demonstrated in the Bandurraga Basin model to the nondimensional equation for water quality (Equation 3.17). What is the finite difference form of the governing equation? Calculate the finite difference solution and compare it with the exact analytical solution; include a plot of the two solutions for comparison purposes. What size time step do you need to use to get a good match with the analytical solution?

6. Suppose you have a container with a volume V of water at an initial temperature of $T(t = 0)$ [°C]. The sides and bottom of the container are insulated (i.e., you can neglect heat losses through the sides and bottom), and the top is losing heat at a rate dictated by Newton's law of convective cooling.

$$q = hA[T(t) - T_\infty], \tag{3.33}$$

Here q is the total heat flux away from the water [W], A is the area available for heat exchange [m^2], $T(t)$ is the temperature of the well-mixed water in the container [°C], T_∞ is the temperature of the air far from the container (toward which the water is cooling), and h is the coefficient of convective cooling [W/m^2°C], which depends on the geometry of the container and the velocity of any air movement across the top of the container, among other factors. At time $t = 0$, you begin adding water to the container at a constant rate and constant temperature; the added water is well mixed, and the volume of water in the container stays the same because the extra goes to an overflow outlet.

(a) Formulate a model for the temperature in the container as a function of time and nondimensionalize your model. Can you solve the model by direct integration? If so, solve the model for nondimensional temperature as a function of nondimensional time.

(b) On the basis of the model you formulated, what is the steady-state temperature of the water in the container? What is the dependence of the steady-state temperature on the initial condition?

(c) Supposing that the initial temperature of the water is $T(t = 0) = 30\,°C$, the input temperature of the water is $T_{in} = 40\,°C$ at a rate of $4\,ml/s$, the far-field temperature $T_\infty = 10\,°C$, the volume and surface area of the container are $V = 4\,l$ and $A = 60\,cm^2$, respectively, and the coefficient of convective cooling $h = 240\,W/m^2°C$. Plot the temperature in the container as a function of time.

CHAPTER 4

The Laplace equation

Chapter summary

The Laplace equation may very well be the single most important equation in mathematical physics. In groundwater studies, it describes the steady flow of groundwater in a domain with no sources or sinks, where all the driving potential is located on the boundaries of the domain. Although it may sound somewhat limited in application, it is a commonly employed conceptual representation of groundwater flow; furthermore, the Laplace equation is a limiting case of several other important field equations we will see in future chapters. In this chapter, we will apply the Laplace equation to understand the configuration of the water table below an idealized region, located somewhere between Heaven and Hell.

4.1 Laplace's equation

In previous chapters, we have been concerned with modeling processes that vary over time. Spatial variability did not enter into our considerations; to the extent that spatial variability was a part of our problems, we glossed over it, using averaged or effective properties to represent the model space. Beginning with this chapter, we will turn our attention to fields that vary in space—for example, the distribution of head in an aquifer, or the distribution of temperature in a steel plate. In this chapter, we will confine ourselves to steady-state processes lacking source/sink terms. These types of problems belong to the class of systems described by the Laplace equation.

The Marquis Pierre-Simon de Laplace (1749–1827) was a famous French mathematician, and author of the *Méchanique Céleste*. Laplace has been credited with making important contributions to many areas of mathematics and physics, including potential theory, the theory of probability, and celestial mechanics, as well as being possibly the first person to propose what we would call a "black hole," by suggesting that a sufficiently massive star might have a gravitational field so great that even light could not escape from its surface. In mathematical physics, the Laplace equation refers to a partial differential

Models and Modeling: An Introduction for Earth and Environmental Scientists, First Edition. Jerry P. Fairley.
© 2017 John Wiley & Sons, Ltd. Published 2017 by John Wiley & Sons, Ltd.
Companion website: www.wiley.com/go/Fairley/Models

equation (PDE) consisting entirely of second derivatives that sum to zero. For example,

$$\frac{\partial^2 H}{\partial x^2} + \frac{\partial^2 H}{\partial y^2} + \frac{\partial^2 H}{\partial z^2} = 0 \tag{4.1}$$

is the 3D Laplace equation in Cartesian[1] coordinates. A more compact notation for Laplace's equation is often used, which is

$$\nabla^2 H = 0. \tag{4.2}$$

This notation not only has the advantage of compactness, but it can also be used when the dimensionality and/or coordinate system is left general (if the dimensionality and or coordinate system is known, it is customary to follow the dependent variable with a list of the independent variables, e.g., $\nabla^2 H(r, \theta, \phi) = 0$ for spherical coordinates).

We will have more to say about the Laplace equation in the near future. For now, we will turn our attention to the problem at hand: a hydrogeologic description of the Elysian Fields.

4.2 The Elysian Fields

These days, it is likely that few people are familiar with the Elysian Fields. In Greek mythology, the Elysian Fields were an infinite tract of grassy plains, beyond the World Encircling Sea, to which heros and mortals related to the gods went after death; they were sometimes called "the Fortunate Isles" or "Isles of the Blessed." There was apparently plenty of food and wine, the weather was always mild, and souls spent most of their time "frolicking."

In *Il Inferno*, the first book of Dante Alighieri's *Commedia Divinia*, Dante paints a somewhat different picture of the Elysian Fields, which he calls "the First Circle of Hell" (the Limbo of the Virtuous Pagans). The "virtuous pagans"—that is, people such as Virgil, Socrates, Plato, and Homer—could be found there, hanging out with all the unbaptized children. The First Circle of Hell was just below the Gate of Hell (the Opportunists) and a quick trip across the River Acheron (the first river of Hell) on Charon's (pronounced "Karen") barge. I realize that most people think of Charon's barge as crossing the River Styx, but Dante tells us the River Styx forms a stinking marsh before the flaming walls of the city of Dis, down at the Fifth Circle of Hell (the Wrathful and Sullen).

On the basis of Dante's account and descriptions from Greek mythology, we can sketch out a kind of composite cartoon of the Elysian Fields (Figure 4.1). Looking at the figure with the eyes of a hydrogeologist, the obvious question that jumps to mind is, what does the potentiometric surface look like beneath the Elysian Fields?

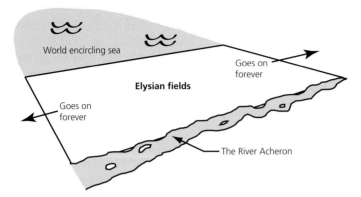

Figure 4.1 Schematic diagram of the Elysian Fields. The representation shown here is a composite of accounts from Greek mythology and *Il Inferno*.

4.3 Model development

Of course, it is a bit of hyperbole to say the "obvious" question to be answered is "what is the configuration of the potentiometric surface beneath the Elysian Fields?" Nonetheless, it will be an instructive question for us to attempt to answer. Fortunately, we have a formalism we may use to approach such a problem: state the quantity that is to be found, list the known data and our assumptions, describe an approach, and carry out the plan.

As in previous chapters, we begin with a statement of that which is to be found.

FIND: The distribution of head in the aquifer underlying the Elysian Fields.

Since we know very little about the system, our list of "knowns" is very short.

KNOWN:

• The Elysian Fields are perfect and infinite in extent.
• The Elysian Fields are bounded by the World Encircling Sea on one side, and the River Acheron on the other.

For this problem, the list of assumptions is necessarily quite long. Because of the number and importance of the assumptions, we will discuss each in detail.

ASSUME:

• *Because the Elysian Fields are part of mythology and are perfect, we can assume they are isotropic and homogeneous.*
This makes perfect sense by the kind of reasoning used in the Middle Ages. A corollary to this assumption is that we must be dealing with a water table

(unconfined) aquifer (you should ask yourself why isotropic and homogeneous conditions imply an unconfined aquifer).

- *The change in the thickness of the aquifer across the domain is a small fraction of the total aquifer thickness; therefore, we can neglect this change.*

  Even though we are dealing with an unconfined aquifer, if the change in the thickness is small compared to the total aquifer thickness, it is reasonable to assume this change is negligible. The problem is then equivalent to that of a confined aquifer, which greatly simplifies the subsequent mathematics. You will have the opportunity to test the validity of this assumption in the problems at the end of this chapter (see Problem 5).

- *The Elysian Fields are infinite in the direction parallel to the River Acheron and the World Encircling Sea.*

  We know the Elysian Fields are both infinite in extent *and* bounded by the River Acheron and the World Encircling Sea (see the knowns listed earlier). The only way we can reconcile these two statements is to postulate that the Elysian Fields are infinite parallel to these two boundaries. Since we are assuming this to be the case, the World Encircling Sea and the River Acheron must continue along either boundary infinitely at the same elevation; otherwise, we would have a logical contradiction, because the River Acheron would have to come from an infinitely high source and flow to an infinite depth. It turns out that the River Acheron is pretty much a marsh or swamp anyway, so it is probably reasonable to assume that it exists everywhere at about the same elevation. We know the River Acheron is a marsh, because Dante tells us so. Speaking in *Il Inferno* of Charon's anger at having to ferry a living human, he says that

  > The steersman of that marsh of ruined souls
  > who wore a wheel of flame around each eye
  > stifled the rage that shook his woolly jowls
  > (Ciardi, 1982, Canto III, lines 94–96)

  We realize an additional benefit from this assumption: since the boundaries are of infinite extent and unchanging, groundwater flow must proceed in a line normal to both the World Encircling Sea and the River Acheron. In other words, we can assume 1D flow.

- *We can neglect tides on the World Encircling Sea.*

  It seems likely that tides on the shores of the World Encircling Sea are small– certainly, they are small in comparison with the thickness of the aquifer we're interested in. Assuming the tides can be neglected allows us to treat the aquifer as steady state. Because we don't need to concern ourselves with changes to the aquifer over time, we can use a much simpler formulation of the governing equation.

- *There is no recharge on the Elysian Fields.*

  I base this assumption on the idea that there is a lot of "frolicking" on the Elysian Fields. Since it seems unlikely that souls would feel much like frolicking in the rain, I think we can reasonably assume no rain (i.e., no recharge). In any case,

Homer tells us in the *Odyssey* that "...no snow is there, nor heavy storm, nor ever rain..."

On the basis of our understanding of the system, we can propose the following approach to defining our model:

APPROACH: Conceptualize the flow as 1D and steady state, with flow from the World Encircling Sea to the River Acheron. Formulate the governing equation using mass balance and solve.

We can also sketch a schematic representation of the model domain (Figure 4.2). The axes and the model domain are shown without the surrounding conceptual information in Figure 4.3.

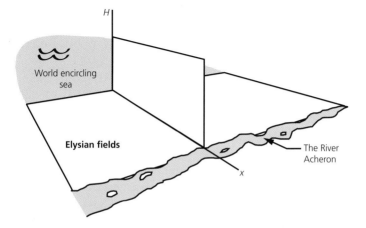

Figure 4.2 Diagram showing the relationship between the 1D model domain and the conceptual representation of the Elysian Fields. The model domain is shown above the ground level of the Elysian Fields for clarity of illustration; in actuality, the model domain is under the ground level.

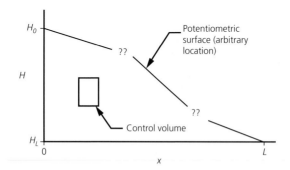

Figure 4.3 The one-dimensional model domain for the Elysian Fields problem. The x-axis of the model domain runs from $x = 0$ at the World Encircling Sea to $x = L$ at the River Acheron. Head is equal to H_0 at the World Encircling Sea, and H_L at the River Acheron. The potentiometric surface is drawn at arbitrary locations–this is the quantity we are trying to find.

4.4 Quantifying the conceptual model

In order to apply mass balance to the conceptualization in Figure 4.2, we will need to define a "control volume," or CV, within the model domain (see Figure 4.3). A CV is an imaginary space that represents any arbitrary location in the model domain. The "control" modifier in front of the "volume" root is added because we can readily understand exactly what goes into, comes out of, and changes within the CV (although we don't actually control it in the usual sense of the word—we just use it to identify changes and fluxes associated with a volume in the model domain). We can conceptualize our CV at any arbitrary location in the model domain, although it shouldn't be drawn in direct contact with any of the domain boundaries. We draw the CV away from the boundaries even though it applies at the boundaries—it's just less confusing to think about it in an arbitrary location away from the boundaries. We can draw a close-up of the control volume (Figure 4.4), label all the fluxes and changes associated with the volume, then use the diagram as an aid to help us develop our balance equation.

As with our previous models, our statement of mass balance is given as:

$$M_{\text{in}} = M_{\text{out}} + \Delta M. \tag{4.3}$$

Because we are looking at a steady-state problem, there will be no changes in the amount of mass stored in the CV, so $\Delta M = 0$. Using Figure 4.4 as a guide, we can see the two remaining terms, M_{in} and M_{out}, can be expanded as:

$$M_{\text{in}} = q_x w b \rho \Delta t, \tag{4.4}$$

$$M_{\text{out}} = q_{x+\Delta x} w b \rho \Delta t, \tag{4.5}$$

where ρ is the density of the water in the aquifer [M/L^3], q is the flux [L/T] evaluated at either x or $x + \Delta x$ as indicated by the subscript, Δt is an arbitrary (but small) unit of time [T], and the product of w and b is the area through which

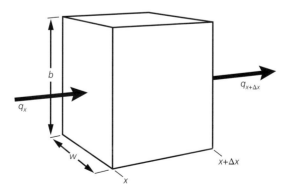

Figure 4.4 Detail of the control volume (CV) for the Elysian Fields problem. The CV is w wide, b tall, and Δx thick in the x-direction. The flux into the CV is labeled q_x, while the flux out of the CV is labeled $q_{x+\Delta x}$. Both fluxes are drawn positive in the positive x-direction.

the flux passes [L^2]. Checking the units, we find that both quantities have units of mass. Next, we substitute Equations 4.4 and 4.5 into Equation 4.3 to get:

$$q_x wb\rho \Delta t = q_{x+\Delta x} wb\rho \Delta t, \tag{4.6}$$

or:

$$q_{x+\Delta x} wb\rho \Delta t - q_x wb\rho \Delta t = 0. \tag{4.7}$$

Dividing both sides of Equation 4.7 by $wb\rho \Delta t \Delta x$ yields:

$$\frac{q_{x+\Delta x} - q_x}{\Delta x} = 0. \tag{4.8}$$

Finally, taking $\lim_{\Delta x \to 0}$ we have

$$\frac{dq}{dx} = 0. \tag{4.9}$$

To proceed beyond this point, we will need to substitute for q in Equation 4.9. We can use Darcy's law,

$$q = -K\frac{dH}{dx} \tag{4.10}$$

as a statement for q in Equation 4.9. Making the substitution yields

$$\frac{d}{dx}\left[-K\frac{dH}{dx}\right] = 0, \tag{4.11}$$

$$-K\frac{d^2H}{dx^2} = 0, \tag{4.12}$$

$$\frac{d^2H}{dx^2} = 0. \tag{4.13}$$

In Equation 4.12, we have removed the $-K$ from under the differentiation operator, which is allowable because it is a constant under our assumptions. In order to have a well-posed problem, we need boundary conditions at $x = 0$ and $x = L$. Since we "know" the heads at these locations and they are assumed not to change over time, we can write the boundary conditions for Equation 4.13 as *specified head* boundary conditions:

$$H(x = 0) = H_0, \tag{4.14}$$

$$H(x = L) = H_L. \tag{4.15}$$

These boundary conditions are shown graphically in Figure 4.3.

4.5 Nondimensionalization

An examination of the boundary conditions (Eqs. 4.14 and 4.15) demonstrates that water levels in the Elysian Fields aquifer are likely to vary between H_0 and

H_L. If we want to nondimensionalize head in the model domain so that it varies between 0 and 1, we can use the normalization

$$\theta = \frac{H(x) - H_L}{H_0 - H_L}. \tag{4.16}$$

Of course, to substitute this expression into the governing equation and the boundary conditions, it is necessary to solve for H and d^2H.

$$H = (H_0 - H_L)\theta + H_L, \tag{4.17}$$

$$dH = (H_0 - H_L)d\theta, \tag{4.18}$$

$$d^2H = (H_0 - H_L)d^2\theta. \tag{4.19}$$

In terms of normalizing the independent variable x, there is a clear length scale on which to nondimensionalize—that is, L, the distance between the World Encircling Sea and the River Acheron, which means our model domain would run from 0 to 1. Normalizing on L and solving for x and dx^2,

$$\xi = \frac{x}{L}, \tag{4.20}$$

$$x = L\xi, \tag{4.21}$$

$$x^2 = L^2\xi^2, \tag{4.22}$$

$$dx = Ld\xi, \tag{4.23}$$

$$dx^2 = L^2d\xi^2. \tag{4.24}$$

Equations 4.17 through 4.24 can be substituted into the governing equation (4.13):

$$\frac{(H_0 - H_L)}{L^2} \frac{d^2\theta}{d\xi^2} = 0, \tag{4.25}$$

$$\frac{d^2\theta}{d\xi^2} = 0. \tag{4.26}$$

Equation 4.26 is the nondimensional governing equation and is subject to the transformed (nondimensional) boundary conditions,

$$\theta(\xi = 0) = 1, \tag{4.27}$$

$$\theta(\xi = 1) = 0. \tag{4.28}$$

4.6 Solving the governing equation

Equation 4.26 belongs to the class of second-order ordinary differential equations that lack a first derivative term. Equations of this type can be solved by direct integration, using a substitution "trick": let $d\theta/d\xi = \chi$; then, integrating both sides with respect to ξ,

$$\frac{d\chi}{d\xi} = 0, \tag{4.29}$$

$$\int \frac{d\chi}{d\xi} d\xi = \int 0 d\xi, \tag{4.30}$$

$$\int d\chi = \int 0 d\xi, \tag{4.31}$$

$$\chi = A_1, \tag{4.32}$$

$$\frac{d\theta}{d\xi} = A_1. \tag{4.33}$$

In Equations 4.32 and 4.33, we have lumped the constants of integration from both integrals together into the constant A_1. Once again, integrating both sides of the equation with respect to ξ,

$$\int \frac{d\theta}{d\xi} d\xi = A_1 \int d\xi, \tag{4.34}$$

$$\int d\theta = A_1 \int d\xi, \tag{4.35}$$

$$\theta = A_1 \xi + A_2. \tag{4.36}$$

We have two constants of integration in Equation 4.36, and we can use the two boundary conditions (Eqs. 4.27 and 4.28) to evaluate them. Substituting $\xi = 0$ and $\theta = 1$, we find $A_2 = 1$. Similarly, substituting $\xi = 1$ and $\theta = 0$ shows that $A_1 = -1$. The complete solution for Equation 4.26 is therefore

$$\theta(\xi) = 1 - \xi. \tag{4.37}$$

4.7 What does it mean?

At this point, we have a solution to our model equation, but we should probably ask ourselves some of the following questions: what does the solution mean? Does the result make sense? How could we use the result? Does the result have any applications in real life? These are important questions, and there are many others like them. You should consider these types of questions for every model you develop.

4.7.1 Considering the solution

The solution to our model, given by Equation 4.37, is a simple linear poten-tiometric surface with a nondimensional slope of -1. A plot of this solution is given in Figure 4.5. Physically, this seems to make reasonable sense. The water table slopes downward from the World Encircling Sea to the River Acheron at a constant rate. With nothing to perturb it along the flowpath—no changes in

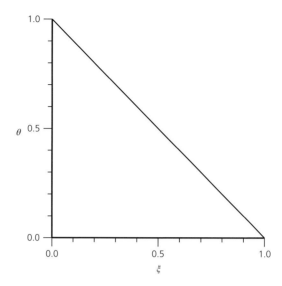

Figure 4.5 Plot of the solution of the Elysian Fields model.

conductivity, no sources for recharge or discharge—there is nothing to cause any variation in the gradient. The solution accords with our physical intuition for how the potentiometric surface should look.

What about mathematically? Does the solution make mathematical sense to us? It is easy enough to verify that the boundary conditions are met. Additionally, Equation 4.33 tells us that the first derivative of our solution is equal to a constant, which accords with the solution; furthermore, we know that constant is -1, which is also clear in our solution and in Figure 4.5. So, mathematically, the solution does seem to make sense.

Beyond the simple tests in the preceding paragraph, we can look at the problem in a more fundamental way. Most people remember from their first course in calculus that the first derivative is the slope of a function; however, many have forgotten that the second derivative of a function can be interpreted as that function's curvature. We can apply this property of the second derivative to the evaluation of our model: it is clear from inspection of the governing equation (4.26) that the curvature of the function θ is zero; in other words, a straight line. We could, therefore, have predicted the solution from knowledge of the governing equation, without performing any integrations at all! Given a solution of zero curvature and boundary conditions of 1 and 0 at $x = 0$ and $x = 1$, respectively, it is clear the solution must be a line between the specified endpoints.

In fact, Laplace's equation (including the 1D version we have been considering) always predicts a linear surface—a line in one dimension, a plane in two dimensions, and a hyperplane in three dimensions. There are a number of other useful consequences of the zero curvature of solutions to Laplace's equation.

For example, because the solution of the equation has zero curvature, there can be no maxima or minima (in mathematical language, "no extrema") within the interior of the model domain. All the extrema must be on the boundaries of the domain. Furthermore, if one considers the problem closely, it becomes clear that any point in the interior of the model domain is the average of the neighboring points—a fact that will be shown quite persuasively when we consider numerical approximations to the Laplace equation in Section 4.8.

4.7.2 The Flux, and its meaning

Since we have a model for the potentiometric surface, we can use that model to calculate the flux from the World Encircling Sea to the River Acheron across the Elysian Fields. Dimensionally, of course, the flux is given as a function of the hydraulic gradient and the hydraulic conductivity as:

$$q = -K\frac{dH}{dx}. \tag{4.38}$$

Unfortunately, we don't know the hydraulic conductivity of the Elysian Fields, so we can't actually calculate the flux between the two boundaries. We can, however, learn something instructive from nondimensionalizing Darcy's law. By substituting Equations 4.18 and 4.23 into Equation 4.38, we find

$$q = -K\frac{(H_0 - H_L)}{L}\frac{d\theta}{d\xi}, \tag{4.39}$$

$$\frac{qL}{K(H_0 - H_L)} = -\frac{d\theta}{d\xi}, \tag{4.40}$$

or

$$v = \frac{q}{-K\frac{(H_L - H_0)}{L}} = -\frac{d\theta}{d\xi}, \tag{4.41}$$

where we define v as the dimensionless flux. It should be clear from Equation 4.41 that the characteristic flux, q_c, is equal to

$$q_c = -K\frac{(H_L - H_0)}{L}, \tag{4.42}$$

which is identical to the "linearized" form of Darcy's law given in Equation 2.7. This linearized Darcy's law is often used to approximate the flux in a system when actual information on the gradient is not available, but the head at the endpoints is known. A little careful thought will show that this approximation is only valid for a linear gradient. In this context, the fact that Laplace's equation predicts a linear gradient and, therefore, a nondimensional flux of 1 makes perfect sense. For any situation in which the gradient deviates from linear, the nondimensional flux v will differ from 1 as dictated by Equation 4.41. This seemingly simple result has subtle and profound implications; as Miyamoto Musashi said in his book *Go Rin No Sho* (*A Book of Five Rings*): "You ought to think deeply about this" (Harris, 1974).

4.7.3 Meanwhile, back in the real world...

This is probably a good point to stop and inquire about my motivation for setting the example problem for Laplace's equation in the mythical Elysian Fields. The reason, obviously, is because the Elysian Fields can be considered "perfect" in a way that no physical location can ever be. As a result, it is a simple matter to claim homogeneous, isotropic materials, and a steady state, infinite aquifer bounded on two sides with infinite, unchanging boundaries. Clearly, such a situation can never occur in the physical world. However, there are certainly locations that approximate these conditions closely enough for Laplace's equation to be a useful model.

To think more carefully about the application of Laplace's equation to the real world, however, consider the following: most groundwater hydrologists with industry experience have learned that it requires three wells to establish the slope of a potentiometric surface. But is this really true? In fact, three wells (yielding the location of the potentiometric surface at three points in space) are sufficient to calculate the orientation of a plane. Three wells will give the orientation of the potentiometric surface of an aquifer, then, provided the potentiometric surface is a plane. Another way of looking at this is that whenever a hydrogeologist uses three wells to establish the slope of a water table, *Laplace's equation is assumed to be a valid model of the aquifer under investigation*. This is hardly a trivial assumption, given all the restrictions placed on Laplace's equation (see the list of assumptions earlier in this chapter).

As we stated earlier, Laplace's equation predicts a surface of zero curvature. We can profitably ask ourselves, then, what conditions lead to a curvature of the potentiometric surface? There are actually four conditions that commonly cause such curvature. These are as follows:

1. The presence of sources or sinks in the domain.
2. Heterogeneity in the porous medium.
3. Transient conditions within the aquifer.
4. A water table aquifer always demonstrates a curved potentiometric surface, unless the water is motionless.

In future chapters, we will consider more general equations that apply to aquifers with these conditions. At present, though, we should ask ourselves how often in real life our hydrogeologic investigations meet with an aquifer that contains no sources or sinks (e.g., no recharge), no heterogeneity, and no transient changes (i.e., is at steady state)? Of course, some sites may conform sufficiently to these conditions that three wells are all that are necessary to determine the slope of the water table. More often, all that is really required is a crude idea of the slope of the potentiometric surface, in which case a three-well approximation may suffice. However, it is easy to construct scenarios where the data collected from three wells are virtually useless for determining the slope of the potentiometric surface, and these scenarios are not outside the realm of reality. It is therefore important to have a clear idea of what is being assumed when one applies a model to an actual field site, and of what the implications of those assumptions are for the situation under consideration.

4.8 Numerical approximation of the second derivative

Earlier (Section 2.8), we proposed a numerical approximation for the first derivative.

$$\frac{d\theta}{d\xi} \approx \frac{\theta_{i+1} - \theta_i}{\Delta\xi}. \tag{4.43}$$

Although this "finite difference" approximation of the first derivative was useful to us in estimating the solutions to equations in earlier chapters, we will need a new operator to approximate the second derivative in Laplace's equation (and other equations). Conceptually, at least, this is a relatively easy thing to do. We know the second derivative is "the derivative of the derivative." As a result, we can find a finite difference operator for the second derivative by taking the finite difference of the first derivative finite difference operator. We can do this as follows:

$$\frac{d^2\theta}{d\xi^2} \approx \frac{\frac{d\theta}{d\xi}|_{\xi+\Delta\xi} - \frac{d\theta}{d\xi}|_{\xi}}{\Delta\xi}, \tag{4.44}$$

$$\frac{d^2\theta}{d\xi^2} \approx \frac{\frac{\theta_{i+1} - \theta_i}{\Delta\xi} - \frac{\theta_i - \theta_{i-1}}{\Delta\xi}}{\Delta\xi}, \tag{4.45}$$

$$\frac{d^2\theta}{d\xi^2} \approx \frac{\theta_{i+1} - 2\theta_i + \theta_{i-1}}{(\Delta\xi)^2}, \tag{4.46}$$

If we apply the approximation in Equation 4.46 to the 1D version of the Laplace equation (Eq. 4.26), we find the finite difference version of the equation is

$$\frac{\theta_{i+1} - 2\theta_i + \theta_{i-1}}{(\Delta\xi)^2} = 0, \tag{4.47}$$

which we can solve to find an expression for θ_i,

$$\theta_i = \frac{\theta_{i+1} + \theta_{i-1}}{2}. \tag{4.48}$$

As was mentioned earlier, the Laplace equation requires the value of the dependent variable at any point to be the average of the neighboring points. This should be clear from Equation 4.48, which shows that θ_i is equal to the arithmetic average of θ_{i+1} and θ_{i-1}.

There are two general approaches for applying Equation 4.48 to the solution of Equation 4.46: iterative methods and direct solution methods. We will briefly examine each of these approaches in the following subsections. However,

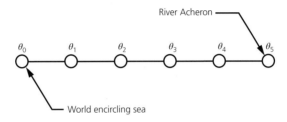

Figure 4.6 Schematic representation of a finite difference grid for the Elysian Fields problem. The 1D grid is discretized into nodes with a $\Delta\xi$ spacing of 0.2. The left-hand boundary node (θ_0) has a constant head of 1, while the right-hand side node (θ_5) has a constant head of 0.

regardless of the method used to solve the governing equation, all finite difference methods require the model domain to be "discretized," or partitioned into a grid of discrete points called "nodes." In the formulation of Equation 4.46 (and hence Eq. 4.48), we have implicitly assumed a constant nodal spacing of $\Delta\xi$. An example grid for the 1D Elysian Fields problem is shown in Figure 4.6. In the figure, a nodal spacing of $\Delta\xi = 0.2$ is used.

4.8.1 Iterative solution methods

The goal of the finite difference method of solution for differential equations is to arrive at estimates of the dependent variable (in this case head, or nondimensional head) at each node in the grid. For the Elysian Fields problem, we know the value of nondimensional head at the two boundaries; iterative application of Equation 4.48 at the unknown nodes will propagate these known conditions across the unknown points. After a sufficient number of iterations across the domain, the change at any given nodal point between the last iteration and the present iteration will fall below some user-specified criterion (known as the "convergence criterion"). When the largest residual is less than this criterion, we say the solution has "converged." In other words, the error between our iterative finite difference solution and the actual (analytical) solution has decreased to a point that we consider acceptable, and we may cease iterating.

Even within the world of iterative solution techniques, there are many different ways to go about solving for the dependent variable. A quick search of the literature will show there are numerous iterative methods for solving the set of equations for θ_i, where $i = 1, N$ and N is the number of nodes in the grid. We will only touch on two of the most basic methods in this chapter; other methods may be found, for example, in Wang and Anderson (1982).

4.8.1.1 Jacobi iteration

The most basic method for iterating to a solution of the finite difference model represented by Figure 4.6 is Jacobi iteration. To implement a Jacobi iteration routine, an initial guess is made for the value of each node in the model grid; typically, a value of zero is used for the initial guess, but convergence can be accelerated if prior information can be used to make the initial guesses closer to the final values. During the first iteration, the finite difference operator (Eq. 4.48) is calculated for each of the nodes, $i = 1, N$, using the known values (from the head boundaries) and the initial guesses. Once a new value has been estimated for all the nodes in the grid, the "residuals" (the absolute values of the differences between the initial guesses and the newly calculated values) are calculated. If the largest residual is smaller than the convergence criterion, the new values are taken as the solution and the calculation is stopped. If, however, one or more of the residuals are greater than the convergence criterion, the old values (the initial guesses for the first iteration, or the values for the previous iteration for later iterations) are replaced by the new values and a new iteration is begun.

4.8.1.2 Gauss–Seidel iteration

The method of Jacobi iteration is reliable but slow; in practice, it is never used. Much better is the method of Gauss–Seidel iteration. This method works similar to Jacobi iteration, but it is faster because the most recently calculated values for the dependent variable are used, rather than waiting to apply the updated values until the next iteration. For example, when calculating the first iteration, we would use the boundary node θ_0 and the initial guess for the interior node θ_2 to calculate the value of θ_1. However, we would use the updated value of θ_1 (the value calculated during the present iteration) along with the initial guess for θ_3 to calculate the new value for θ_2. We continue to calculate our updated estimates at each node in this way, always using the most recent value available for any node, until the residuals for all the nodes fall below the convergence criterion. Using the most up-to-date estimates for our calculation moves information more quickly from the boundaries to the interior nodes, which speeds convergence considerably in comparison to Jacobi iteration.

4.8.2 Direct solution methods

In comparison to iterative methods, direct solution methods use the fact that the equations for each node (i.e., Eq. 4.46, with $i = 1, N$, where N is the number of nodes in the grid), when taken together, form a system of N equations with N unknowns. This system can then be solved simultaneously by matrix methods.

Using the example grid for the Elysian Fields problem shown in Figure 4.6, we can write four equations, one for each of the interior nodes, as:

$$-\frac{1}{2}\theta_0 + \theta_1 - \frac{1}{2}\theta_2 = 0, \tag{4.49}$$

$$-\frac{1}{2}\theta_1 + \theta_2 - \frac{1}{2}\theta_3 = 0, \tag{4.50}$$

$$-\frac{1}{2}\theta_2 + \theta_3 - \frac{1}{2}\theta_4 = 0, \tag{4.51}$$

$$-\frac{1}{2}\theta_3 + \theta_4 - \frac{1}{2}\theta_5 = 0. \tag{4.52}$$

In Equation 4.49, θ_0 has a fixed value of 1 because it represents the boundary condition at $\xi = 0$. Similarly, θ_5 in Equation 4.52 represents the boundary condition at $\xi = 1$, and has a fixed value of 0. We can move the known values (the boundary conditions) to the right-hand sides of the respective equations, and write the entire system as a matrix equation:

$$
\begin{bmatrix}
1 & -\frac{1}{2} & 0 & 0 \\
-\frac{1}{2} & 1 & -\frac{1}{2} & 0 \\
0 & -\frac{1}{2} & 1 & -\frac{1}{2} \\
0 & 0 & -\frac{1}{2} & 1
\end{bmatrix}
\begin{bmatrix}
\theta_1 \\
\theta_2 \\
\theta_3 \\
\theta_4
\end{bmatrix}
=
\begin{bmatrix}
\frac{1}{2} \\
0 \\
0 \\
0
\end{bmatrix}.
\tag{4.53}
$$

A great deal of effort has been expended on methods to solve systems of equations written as matrices. Gaussian row reduction is a common method taught in most linear algebra courses; however, there are a great many methods that work faster for particular types of matrices. For a review of some popular methods, the interested reader is referred to Press et al. (1992).

In the present instance, the coefficient matrix in Equation 4.53 is a special type that is known as a "tridiagonal matrix." Tridiagonal matrices (matrices in which the main diagonal and the two flanking diagonals of the coefficient matrix are filled, while all other elements are zeros) are quickly and easily solved by the well-known Thomas algorithm. The Thomas algorithm works well for 1D finite difference problems, which are always tridiagonal; for higher dimensional problems, the reader may find it most convenient to use a library routine matrix solver. There is a bare outline of a method for applying the Thomas algorithm to higher dimensional problems in Appendix C, and this method is described in more detail in Patankar (1980). Alternatively, consult Press et al. (1992) or other references for appropriate solution methods.

4.9 Conclusions

In this chapter, we have examined the Laplace equation, which describes steady-state groundwater flow in the absence of any source/sink terms for a homogeneous, isotropic aquifer system. Laplace's equation is a commonly used model, not only for groundwater flow, but also for most potential fields (e.g., heat, species concentration, gravitational potential, electrical charge). As an example, a simplified "one-dimensional Laplace equation" was used to find the potentiometric surface of the aquifer underlying the Elysian Fields, a mythical region described in Greek mythology, and in literature by Dante Aleghieri in his epic poem *Il Inferno*. The solution to the governing equation for the Elysian Fields demonstrated the potentiometric surface must be linear, falling from a high at the World Encircling Sea to a low at the River Acheron. This linear surface is an important feature of solutions to Laplace's equation. We learned that no solution of Laplace's equation can demonstrate extrema internal to the model domain; all extrema must be on the boundaries of the problem. Furthermore, Laplace's equation predicts that every point in the problem domain is the average of the neighboring points, and this interpretation of the equation was born out by the form of the finite difference operator for the second derivative. Strictly speaking, the so-called 1D Laplace equation examined in this chapter is not a true Laplace equation, because Laplace's equation is a PDE, the solutions of which are harmonic functions. Despite this, the simplified version used in this chapter provides a good illustration of many of the features of Laplace's equation.

From a practical standpoint, we have seen that some of the procedures commonly followed in, for example, environmental consulting companies, derive from the assumption that Laplace's equation adequately describes the subsurface head distribution. This may or may not be a good assumption for a given aquifer, but it is important to evaluate the assumption before relying on it in a field setting.

As was noted in the introductory section to this chapter, the Laplace equation is one of the most important equations of mathematical physics. Time spent understanding this fundamental equation will be repaid manyfold; this is particularly true because Laplace's equation is a special case of many of the equations we will be looking at in subsequent chapters.

4.10 Problems

1. Nondimensionalize the following field equations:

   (a)

   $$\frac{\partial^2 H}{\partial x^2} + \frac{\partial^2 H}{\partial y^2} = 0, \tag{4.54}$$

   $$H(x = 0, y) = H_0, \tag{4.55}$$

   $$H(x = L, y) = H_0, \tag{4.56}$$

   $$H(x, y = 0) = H_0, \tag{4.57}$$

   $$H(x, y = L) = H_L, \tag{4.58}$$

   (b)

   $$\frac{\partial^2 T}{\partial x^2} + \frac{\partial^2 T}{\partial z^2} = 0, \tag{4.59}$$

   $$T(x = 0, z) = T_1, \tag{4.60}$$

   $$T(x = L, z) = T_2, \tag{4.61}$$

   $$T(x, z = 0) = T_1, \tag{4.62}$$

   $$T(x, z = L) = T_3, \tag{4.63}$$

   with $T_1 < T_2 < T_3$. HINT: You will need to define a dimensionless constant in this problem. Perhaps you could call it θ_2 or similar.

2. If $H_0 = 0$ masl, $H_L = -50$ masl, $L = 10$ km, and $K = 10^{-5}$ m/s, what is
the flux (in m/s) across the Elysian Fields? What is the total discharge from
the World Encircling Sea to the River Acheron? You may assume the aquifer
thickness is 500 m. Explain the value you estimated for the total discharge
across the Elysian Fields. Does it matter what value you use for the thickness
of the aquifer?

3. Use Gauss–Seidel iteration to find the values of θ_1, θ_2, θ_3, and θ_4 in the Elysian
Fields problem with a convergence criterion of 0.1 and an initial guess for
the nondimensional-head at each point of 0. How many iterations did it
require to reach convergence? How does the numerical solution compare to
the analytical solution (Eq. 4.37)?

4. Use a prepackaged software routine, a calculator with a matrix solver, or write
your own version of the Thomas algorithm to solve Equation 4.53 (if you
program in Fortran, you may use the subroutine given in Appendix C). How
does your solution compare to the solution obtained by Gauss–Seidel iteration? How does it compare to the analytical solution given by Equation 4.37?

5. The derivation of the governing equation for the Elysian Fields is general up
to Equation 4.7. Dividing both sides of Equation 4.7 by $\Delta t \Delta x$ gives

$$w\rho \frac{b(x + \Delta x)q_{x+\Delta x} - b(x)q_x}{\Delta x} = 0. \tag{4.64}$$

Here, we have retained b as a function of x because, in order to model a
water table aquifer, b must be allowed to vary with distance across the aquifer
(dividing out b, as we did in Equation 4.8, is only valid for a confined aquifer
where b is a constant). Dividing by $w\rho$ and taking $\lim_{\Delta x \to 0}$,

$$\frac{db(x)q}{dx} = 0. \tag{4.65}$$

As before, we can substitute Darcy's law for q. Making the substitution and
dividing both sides by $-K$ gives

$$\frac{d}{dx}\left[b(x)\frac{dH}{dx}\right] = 0. \tag{4.66}$$

Equation 4.66 is nonlinear because the thickness of the aquifer, b, changes as
a function of the gradient. Nonlinear equations are generally difficult to solve,
and hydrogeologists therefore resort to numerical methods to approximate a
solution. In this case, however, we can take advantage of the fact that head can
be referenced to any arbitrary datum to find a clever way to solve our problem.
If we reference head to the bottom of the aquifer, then the thickness of the
aquifer (b) is approximately equal to the head in the aquifer, so Equation 4.66
can be written:

$$\frac{d}{dx}\left[H\frac{dH}{dx}\right] = 0. \tag{4.67}$$

This approximation is really only valid when the change in thickness of the aquifer is small in comparison to the total thickness of the aquifer, so that we can neglect the small vertical gradients that arise as the aquifer thickness changes. We can now make use of the fact that

$$\frac{d}{dx}[H^2] = 2H\frac{dH}{dx}.$$
(4.68)

Substituting Equation 4.68 into Equation 4.67, we have

$$\frac{d^2h}{dx^2} = 0,$$
(4.69)

where

$$h = \frac{1}{2}H^2.$$
(4.70)

(You should verify for yourself that this transformation works.) Equation 4.69, obtained using Equation 4.70 and the assumption that vertical gradients can be neglected, is known as the *Dupuit–Forchheimer equation*. Equation 4.70 can also be applied to the boundary conditions (Eqs. 4.14 and 4.15) to obtain:

$$h(x = 0) = h_0,$$
(4.71)

$$h(x = L) = h_L.$$
(4.72)

(a) Solve Equation 4.69 using the direct integration technique demonstrated in this chapter. You will need to apply the boundary conditions (Eqs. 4.71 and 4.72) to find the constants of integration.

(b) Back-transform your solution to Equation 4.69 using Equation 4.70 to obtain a solution in terms of H.

(c) Re-dimensionalize the solution of Equation 4.26 (Eq. 4.37), and plot it together on the same set of axes with your back-transformed solution to Equation 4.69. How different is the solution for the water table aquifer from the solution given by Equation 4.37? Is the error significant enough to justify the extra effort to obtain the water table solution?

(d) Use your back-transformed solution to calculate the flux through the water table aquifer; you should use the parameter values from Problem 2 to calculate the flux. How does the value you calculate compare to the flux you calculated in Problem 2? Is the difference great enough to justify the effort involved in calculating the water table solution?

6. Laplace's equation is given in radial (r, ϕ) coordinates as:

$$\frac{\partial}{\partial r}\left[r\frac{\partial\theta}{\partial r}\right] + \frac{1}{r}\frac{\partial^2\theta}{\partial\phi^2} = 0.$$
(4.73)

For a circular domain $r \leq 1$ and a boundary condition,

$$\theta(r = 1, \phi) = \sin\phi,$$
(4.74)

can you use your knowledge of Laplace's equation to guess the distribution of θ in the model domain? HINT: Try plotting or drawing a sketch of the model domain and marking points of known θ on the boundary, then contour the interior of the model domain. Remember two important constraints of solutions to the Laplace equation: the surface is linear, and there are no interior extrema.

Note

1 You may wonder why "Cartesian" is always capitalized, while radial, spherical, or curva-linear coordinates are never capitalized. The reason is that Cartesian coordinates are named for their originator, René Descartes (1596–1650).

References

Ciardi, J. (1982) *The Inferno, by Dante Aleghieri; A Verse Rendering for the Modern Reader*, New American Library, New York.

Harris, V. (1974) *A Book of Five Rings, by Miyamoto Musashi*, The Overlook Press, Peter Mayer Publishers, Inc., New York.

Patankar, S.V. (1980) *Numerical Heat Transfer and Fluid Flow*, Routledge, Taylor & Francis Group, New York.

Press, W.H., Flannery, B.P., Teukolsky, S.A., and Vetterling, W.T. (1992) *Numerical Recipies in Fortran 77: The Art of Scientific Computing*, Cambridge University Press, Cambridge.

Wang, H.F. and Anderson, M.P. (1982) *Introduction to Groundwater Modeling: Finite Difference and Finite Element Methods*, Academic Press, San Diego.

CHAPTER 5

The Poisson equation

Chapter summary

In this chapter, we will consider the Poisson equation. Poisson's equation is another PDE in the same family (elliptic PDEs) as Laplace's equation, and is in fact an extension or generalization of the Laplace equation. Like Laplace's equation, the derivatives in Poisson's equation are entirely of second-order and are usually assumed to apply to a spatial domain. Also like Laplace's equation, Poisson's equation describes a steady-state potential field. Unlike Laplace's equation, however, the derivatives in Poisson's equation do not (in general) sum to zero. Instead, the derivatives sum to either a constant or a term that is a function of space only (not of time); this term represents a source/sink, such as infiltration to an aquifer from precipitation, discharge from evapotranspiration, recharge from a lake or stream, and so on. Poisson's equation is therefore the nonhomogeneous equivalent of Laplace's equation, which is a special case of Poisson's equation (the homogeneous, or degenerate, case). In the process of applying Poisson's equation to an example problem, we will learn about a broader range of boundary conditions and consider several points of practical importance, such as how can a nonhomogeneous PDE be nondimensionalized, when can a term in an equation be neglected, and how to construct 2D and 3D finite difference operators.

5.1 Poisson's equation

As with Laplace's equation, the Poisson equation is one of the important equations of mathematical physics, where it is commonly used to represent the potential field that results from some sink/source term. The Poisson equation is perhaps most familiar from its role in electrostatics, where it describes the electrical potential field that results from a known charge distribution. In groundwater hydrology, Poisson's equation describes the potentiometric surface in an aquifer that results from some combination of forcing on the domain boundaries and source/sink terms internal to the domain. In three-dimensions (Cartesian coordinates), Poisson's equation is written:

$$\frac{\partial^2 H}{\partial x^2} + \frac{\partial^2 H}{\partial y^2} + \frac{\partial^2 H}{\partial z^2} = f(x, y, z). \tag{5.1}$$

Models and Modeling: An Introduction for Earth and Environmental Scientists, First Edition. Jerry P. Fairley.
© 2017 John Wiley & Sons, Ltd. Published 2017 by John Wiley & Sons, Ltd.
Companion website: www.wiley.com/go/Fairley/Models

Poisson's equation can also be written with the shorthand "grad" notation:

$$\nabla^2 H(x, y, z) = f(x, y, z). \tag{5.2}$$

In Equations 5.1 and 5.2, the nonhomogeneous term $f(x, y, z)$ is either a source or a sink (i.e., a recharge or discharge term) internal to the model domain; for example, $f(x, y, z)$ may represent recharge to the aquifer from precipitation at the land surface, recharge from an infiltration pond overlying the aquifer, or discharge from the aquifer due to plants taking up groundwater and transpiring it to the atmosphere. The close correspondence in form between Poisson's equation and Laplace's equation is no accident: Laplace's equation is the "degenerate" case of Poisson's equation, in which the source/sink term $f(x, y, z) = 0$. Thus, some of the intuition we developed about the potentiometric distribution in connection with Laplace's equation will be helpful to us in understanding what behavior to expect from Poisson's equation, and how to apply it to problems of interest. To begin with, we will describe the domain of our example problem, which is set on a small, but well-known, island in the San Francisco Bay.

5.2 Alcatraz island

Most people are probably familiar with the beautiful City of San Francisco—the jewel of the US west coast, and perhaps the most European of all American cities. San Francisco sits at the northern tip of the San Francisco peninsula; its skyline is dominated by the iconic TransAmerica pyramid (and a really ugly cell phone tower). More famous yet is the Golden Gate bridge, once the longest suspension bridge in the world (and now the second longest in the United States, after the Verrazano Narrows bridge, and the eighth longest suspension bridge in the world). Less familiar to many people are the myriad small islands that dot the San Francisco Bay; for example, Yerba Buena island ("yerba buena" is Spanish for "good herb," more commonly called mint), which is the current home of the Yerba Buena Naval Training Station, or Treasure island, which is an artificial island that was constructed as the centerpiece of the 1939 World's Fair. There is one island in the San Francisco Bay, however, that almost everyone has heard of (Americans, at least), and many people have even visited: Alcatraz island, often known colloquially as "The Rock."

5.2.1 Early history of Alcatraz island

Alcatraz island was originally a lighthouse, before being turned into a military fortress. Sometime before 1906 (the year of the great San Francisco earthquake), it was turned into a military prison; in 1934, it was transferred to the Justice Department and modernized for use as a civilian prison. It operated as a civilian prison until it was closed by President J.F. Kennedy on March 21, 1963. The prison was closed primarily because it was too expensive to operate, costing around $10/day per prisoner (in 1960 dollars), as compared to about $3/day per prisoner for a land-based prison. Why was Alcatraz prison so expensive to operate? There are no natural sources of water on the island; therefore, all water must be brought over from the peninsula by boat, which is an expensive proposition.[1]

Many people are aware of the celebrated escape attempt of Frank Morris and John and Clarence Anglin, thanks to the well-known Clint Eastwood movie, appropriately titled *Escape from Alcatraz*. (The Anglin brothers and Frank Morris are officially listed as "drowned," although no bodies were ever recovered.) It is widely believed that it is impossible to swim from Alcatraz island to San Francisco (or anywhere else, for that matter), and that anyone foolish enough to try to do so will be drowned or eaten by sharks. However, I can say with absolute certainty that it is perfectly possible to swim from Alcatraz to Aquatic Park in San Francisco, provided the attempt is timed correctly to avoid being swept out to sea by the tides. I can say this with such assurance because I myself made that swim, in the early morning hours of September 4, 1999.

5.2.2 The American Indian occupation

Our story today begins in 1971, about 18 months after the island had been occupied by a group of American Indians that referred to themselves as "Native Americans of Different Tribes." Their occupation was based on the claim that the 1868 Fort Laramie treaty between the US government and the Sioux Tribes gave Native Americans acquisition rights to all retired, abandoned, or unused Federal lands that were originally acquired by the government from Native Americans (so, basically—everything).

As was the case when the island housed an operating prison, all water for the occupation was brought from the mainland by boat. After about 18 months, the government decided it had had enough—perhaps as a result of the unwanted publicity generated by a high-profile article in *Time* magazine—and decided the easiest way to end the occupation was to shut off the supply of water.

Picture yourself as a member of the group "Native Americans of Different Tribes." You and your fellow protestors are sitting around, wondering what you are going to do without any water. After a while, though, you begin to have some hazy recollections of an apparently useless groundwater modeling class you took...back in the day. Thinking this might be a way to score some points with your fellow "occupants," you mention that it should be possible to dig a well and have "all the water we need." Your pronouncement is greeted with great joy, and many of your friends run off to find spoons with which to dig a well (after all, it worked for the Abbé Faria in Alexander Dumas's classic tale *The Count of Monte Cristo*[2]). One of the leaders of the occupation is skeptical, however, and immediately begins asking pointed questions. Can it really be done? Where would the water come from? Would the water be salty, on this island in the middle of a brackish bay? Where on the island should the well be located?

You are feeling increasingly uncomfortable under this scrutiny, and now there are murmurs from the onlookers. You could run and jump in the bay, but you believe (mistakenly, as it turns out) it is impossible to swim away from Alcatraz. Instead, you dredge deep within your memory for a way to answer the leader's questions. Clearly, what is required is a model of groundwater flow beneath Alcatraz island. If only you had paid more attention during class! But it's too late for that now, so you ask for a piece of paper and a pencil and begin writing…

5.3 Understanding the problem

Of course, by this time we have a pretty good idea of what our hero should be writing. We will need a clear and succinct statement of that which is to be found, things that are known and things that are assumed about our model domain, and some kind of approach to solving the problem. Once these things are in order, we can go about carrying out our plan. As often happens, it turns out that carrying out the plan is (relatively) the easy part. Understanding how to devise the conceptual model, figuring out which assumptions are reasonable and which aren't, and trying to come up with a model formulation that is in some way physically meaningful while still being analytically tractable…these are the true test of a modeler's craft.

5.3.1 Developing an approach

Starting off with a statement of our objective for the model:

FIND: The distribution of head in the aquifer underlying Alcatraz island, as a function of the rate of recharge (from precipitation), the hydraulic conductivity of the aquifer, and so on.

Once we know the distribution of head in the aquifer, we can use it to calculate the depth we have to dig to reach water (although we may have to correct for the curvature of a water table aquifer; see Problem 5, Chapter 4). We should also be able to guess where the most propitious place is to site a well. In addition, we can hope to gain some insight on how much water might be available if we can estimate the flux discharging from the island into the bay. Similar to the Elysian Fields problem (Chapter 4), we actually know relatively little about the model domain.

KNOWN:

- The geometry of the island (e.g., its size and shape and the elevation of the island surface above the surface of the surrounding bay)
- The approximate amount and timing of rainfall

Also similar to the Elysian Fields problem, we are going to be forced to assume a great many things. In the case of the Elysian Fields, though, we had to assume things because we knew almost nothing about the domain, mostly because it is a mythical place, and thus not well characterized, hydrologically speaking. Here, we are trying to model a real place; at least in theory, we should have more information about our site. We have two issues, however, in the present problem: (i) in effect, we are developing a preliminary model of a site that will guide future characterization efforts (i.e., the placement of a well). In such a situation, it is common to have little or no hard data—we are primarily trying to get an idea of how the system functions. In addition, (ii) the site we have chosen to investigate is actually quite complex; as a result, we will need to make dramatic simplifications to arrive at an analytically tractable representation. Once we have a clearer understanding of how the system works, what aspects are important and which can be neglected, and so on, we can move on to a more complex model of the system (perhaps a finite difference model). Keeping this in mind, we can begin our list of assumptions.

ASSUME:

1. *Homogeneous and isotropic medium.* This should be a familiar assumption, and is one that we used in developing the Laplace equation in connection with the Elysian Fields problem. Unlike that model, however, the hydraulic conductivity will not simply cancel out in the present problem, but rather will play in important role in understanding the distribution of head in the aquifer below Alcatraz island.

   The assumption of homogeneous and isotropic materials is common to most analytical models, not only in groundwater hydrology but also in heat transfer and many other disciplines. The fact that it is common, however, does not necessarily mean it is a good assumption. We will have more to say about this assumption later in the chapter.

2. *Constant density fluid.* The important point for this assumption is that we will not concern ourselves with density differences between salt (or brackish) water and freshwater. The impact of neglecting these density differences is probably minor in terms of the model boundary conditions but it is major in terms of understanding the shape of the water table and the potability of the water in the aquifer. In fact, freshwater tends to form a lens-shaped body that floats on a larger body of saltwater; furthermore, the edges of the freshwater body will have a gradient of salt as salt diffuses from higher to lower concentrations. These effects are beyond our ability to represent in such a simple model, so we will neglect them—but not forget they are there.

3. *Changes in thickness of the aquifer are negligible.* This assumption, too, should be familiar from Chapter 4. It amounts to reducing the complexity of the problem from a water table aquifer to a confined aquifer, which is a considerable simplification of the mathematics. In the case of the Elysian

Fields, this was probably a very good assumption; in the present case, it is perhaps not so good. Once again, you will have the opportunity to test the error of this assumption in the problems at the end of this chapter.

4. *Neglect tides and consider recharge at an average and constant rate.* This assumption amounts to a decision to consider the system to be at steady state. Of course, changing tides and time-varying recharge can be considered in a model—even in an analytical model—but we have not yet developed sufficient complexity of mathematical expression to include these factors. For the time being, an assumption of steady state is probably reasonable.

Finally, we are in a position to write our approach to solving the problem.

APPROACH: Conceptualize the aquifer as a steady-state system, surrounded on all sides by boundaries of constant head (about which, however, more later) and recharged at a steady, average rate by precipitation that falls equally on all points of the island. Use mass balance to formulate the governing equation for the problem and solve.

5.3.2 Questions about coordinates

At this point, we have developed a reasonable understanding of the problem, but we are not yet ready to begin putting our plan into practice, because there are still some outstanding questions we must answer. In particular, we need to decide in what coordinate system we will render our model.

With a numerical model, it is relatively easy to build irregular features into a model grid, so we could develop a model that honors the actual shape of Alcatraz island (within reason). In an analytical model, irregular domains are not easy to accommodate. Given the simplicity of the model, however, the exact shape of the island may not be critical to obtaining an idea of how the island functions, hydrologically. When working with analytical models, it is usually best to approximate the model domain as a simple geometric figure; for example, a square, a rectangle, or a circle. We can guess in the present situation that a circle or a rectangle will probably be the most likely shape for our domain, but the question is, which?

A circular domain is especially attractive for modeling Alcatraz island. Since we are already assuming a homogeneous and isotropic porous medium, we could, perhaps, take advantage of radial symmetry to represent the problem as 1D flow (see Figure 5.1). It is always appealing to reduce the dimensionality of a problem, because the cost (in terms of difficulty of analytical solution or numerical complexity, computational time, and the modeler's time) increases roughly as the power of the number of dimensions. But can we actually use radial symmetry to reduce the dimensionality of the Alcatraz problem? Inspection of a map of Alcatraz (Figure 5.2) shows that, in contrast to the wishful thinking embodied in Figure 5.1, a radial model is probably not appropriate. Instead, we

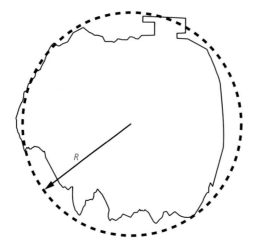

Figure 5.1 Schematic diagram of a radial domain for the Alcatraz model. The model domain is shown as a circle (dotted line) of radius R. As discussed in the text, this model is appealing, but it does not fit well with the actual site geometry.

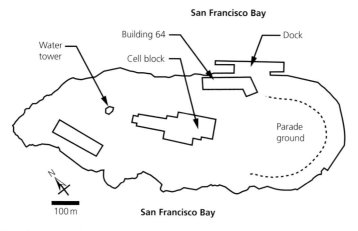

Figure 5.2 Scale map of Alcatraz island, showing selected buildings. Some buildings are omitted for clarity of illustration.

will propose a 2D rectangular domain and Cartesian coordinates. The rectangular model boundary, along with two possible coordinate axes systems, is shown in Figure 5.3.

5.3.3 A digression about boundary conditions

Both sets of coordinate axes displayed in Figure 5.3 possess qualities that recommend them, as well as qualities that argue against their adoption. The axes (x_1, y_1) have the advantage of simplicity: it is intuitive to have the origin of the coordinate system at one corner of the model domain boundary. Furthermore, the four

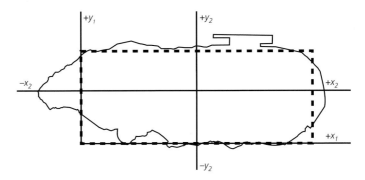

Figure 5.3 Drawing of a rectangular model boundary (dotted line), superposed over an outline of Alcatraz island (map outline drawn to scale). Along with the proposed model boundary, two possible sets of coordinate axes are shown; the first set (x_1, y_1) is centered on the lower left-hand corner of the rectangular boundary. The origin of the second possible set of axes (x_2, y_2) is at the center of the model domain.

boundaries of the model domain, all of which are constant head boundaries, are located at (mathematically) convenient places: two of the boundaries are located at $x = 0$ and $y = 0$, while the other two are at specific values of x and y. At least at first glance, the coordinate system (x_2, y_2) is less convenient, chiefly because the constant head boundaries are located at both positive and negative values of x and y (i.e., none of the boundaries are located at $x = 0$ or $y = 0$). To understand why the (x_2, y_2) coordinate system may be superior to the (x_1, y_1) axes, we need to learn a little more about types of boundary conditions.

So far, we have made use of only one kind of boundary condition—specified (or "constant") head boundaries. In fact, there are (at least) three kinds of boundary conditions that can be employed in groundwater modeling. These boundary conditions are known, appropriately, as boundary conditions of the first kind, second kind, and third kind, and they are described in the following subsections.

5.3.3.1 Boundary conditions of the first kind

Boundary conditions of the first kind are also known as *Dirichlet* or *specified potential* boundary conditions. In groundwater hydrology, they are often termed *specified head*, or even *constant head* (if they do not vary over time). The general case of a boundary condition of the first kind is written mathematically as:

$$H(x = 0, y, t) = f(x = 0, y, t). \tag{5.3}$$

In Equation 5.3, we have written the condition as it would apply to a boundary in a Cartisian coordinate system at $x = 0$; similar conditions could be specified for $x = L$, $y = 0$, $y = L$, or any other boundary of a model domain. Furthermore, we have specified in our condition that the potential on the boundary may vary as a function of both space (at $x = 0$, but in the direction parallel to the y-axis) and time. If the function on the right-hand side of Equation 5.3 is either a constant or a function of space only, hydrogeologists will generally refer to it as a *constant head* boundary.

The specified potential boundary is probably the most intuitive and commonly used boundary condition in groundwater modeling, and it is the boundary condition that we used in our discussion of the Laplace equation. It is not, however, the only boundary condition available to us; equally useful are boundary conditions of the second kind.

5.3.3.2 Boundary conditions of the second kind

Boundary conditions of the second kind are often called *Neumann* boundary conditions. In a boundary condition of the second kind, the derivative (gradient) of the dependent variable is specified normal to the boundary. For example, in the case of a specified gradient on the boundary $x = 0$,

$$\frac{dH}{dx}(x = 0, y, t) = f(x = 0, y, t). \tag{5.4}$$

A term often used in groundwater modeling is a "specified flux" boundary condition. This term for a boundary condition of the second kind arises because a modeler may wish to allow some known or assumed flux to cross a boundary into the model domain; for instance, a flux from a lake or river. The specification of a flux boundary is made mathematically as:

$$-K\frac{dH}{dx}(x = 0, y, t) = q(x = 0, y, t), \tag{5.5}$$

$$\frac{dH}{dx}(x = 0, y, t) = -\frac{q(x = 0, y, t)}{K}. \tag{5.6}$$

A special case of the second kind of boundary condition, often used in groundwater modeling, is the *no-flow* boundary. No-flow boundaries are derived by the obvious specialization of Equation 5.6, where the flux is prescribed as zero:

$$\frac{dH}{dx}(x = 0, y, t) = 0. \tag{5.7}$$

In heat transfer applications, the no-flow boundary is commonly referred to as an "adiabatic" boundary (i.e., an insulated boundary, or a boundary where the flux of heat is zero).

5.3.3.3 Boundary conditions of the third kind

Boundary conditions of the third kind (sometimes referred to as *Robin* conditions) are rarely used in groundwater modeling—at least, explicitly. Even so, it is very important that modelers become familiar with them, because boundary conditions of the third kind are a generalization of boundary conditions of the first and second kinds; that is, boundary conditions of the third kind reduce to boundary conditions of either the first or second kind in certain special cases (discussed in the following text).

A boundary condition of the third kind is more easily visualized in the context of heat transfer than groundwater flow. In heat transfer, boundary conditions of the third kind are sometimes referred to as *convection* (or *convection/conduction*) boundary conditions. A boundary condition of this type is written:

$$-K_T \frac{dT}{dx}(x = 0, y, t) = h[T(x = 0, y, t) - T_\infty]. \tag{5.8}$$

The reason this type of boundary condition is called a "convection/conduction" boundary is apparent from inspection of Equation 5.8. On the left-hand side of the equation, heat is being transferred through solid material by conduction at a rate that is proportional to the gradient and the thermal conductivity $(K_T, [ML/T^3\Theta])$.[3] On the right-hand side, heat is being carried away from the boundary by a fluid at a temperature (far from the boundary) of T_∞. The driving potential for heat transfer on the fluid side is the difference between the temperature of the solid surface $(T(x = 0, y, t))$ and the far-field fluid temperature (T_∞). The parameter h $[M/T^3\Theta]$ is a proportionality constant called *Newton's coefficient of convective cooling*. These two rates (the rate at which heat is leaving the solid and the rate at which heat is being carried away by the fluid) must be equal at the boundary.

To see why the boundary conditions of the first and second kinds are special cases of boundary conditions of the third kind, we can nondimensionalize Equation 5.8 using the following nondimensional variables:

$$\theta = \frac{T - T_\infty}{T_c}, \tag{5.9}$$

$$T = T_c\theta + T_\infty, \tag{5.10}$$

$$dT = T_c d\theta, \tag{5.11}$$

and

$$\xi = \frac{x}{L_c}, \tag{5.12}$$

$$x = L_c\xi, \tag{5.13}$$

$$dx = L_c d\xi, \tag{5.14}$$

where T_c is a characteristic temperature for the system and L_c is a characteristic length scale for the system. Substituting into Equation 5.8 and rearranging gives

$$\frac{d\theta}{d\xi}(\xi = 0, \zeta, \tau) + \beta\theta(\xi = 0, \zeta, \tau) = 0, \tag{5.15}$$

where $\beta = hL_c/K_T$ is a dimensionless parameter known at the *Biot number* (pronounced "BEE-oh"), often abbreviated as "Bi" or similar. The Biot number quantifies the relative efficiency of heat transfer in the fluid by convection to the efficiency of conductive heat transfer in the solid. For large values of the Biot number (e.g., >10), convection dominates over conduction; the fluid can carry away more heat than the solid can provide by conduction, and Equation 5.15 reduces to

$$\theta(\xi = 0, \zeta, \tau) = 0, \tag{5.16}$$

which is a constant potential boundary (a boundary condition of the first kind). If the Biot number is small, heat is transferred through the solid at a greater rate than the fluid can carry it away, and Equation 5.15 becomes

$$\frac{d\theta}{d\xi}(\xi = 0, \zeta, \tau) = 0, \tag{5.17}$$

which is an adiabatic (insulated) boundary condition (a boundary condition of the second kind). In fact, Equation 5.15 is a slightly simplified version of the complete boundary condition of the third kind; the most general statement of a boundary condition of the third kind is:

$$K_T \frac{dT}{dx}(x = 0, y, t) + h[T(x = 0, y, t) - T_\infty] = f(x = 0, y, t). \tag{5.18}$$

Equation 5.18 is the nonhomogeneous version of Equation 5.8, which is the homogeneous version of a boundary condition of the third kind. You can verify for yourself that this general statement of the boundary condition of the third kind reduces to the general statements for boundary conditions of the first and second kinds for appropriate values of the Biot number.

In addition to the fact that the first and second kinds of boundary conditions are special cases of the third kind, there is another more important reason why groundwater modelers should understand boundary conditions of the third kind. It is a common occurrence for a modeler to develop an analytical equation that is beyond the modeler's ability (or patience) to solve, but which may in fact be solvable. In this case, modelers will generally resort to seeking a solution in one of the many published compendia of solutions; for example, Carslaw and Jaeger (1959) or Özişik (1993). The authors of these compendia are likely to write the boundary conditions in the form of Equation 5.18, in which case it is up to the modeler to reduce the solution to the appropriate special case by a considered choice of the Biot number.

5.3.4 Considerations of symmetry

Now that we have a better understanding of the boundary conditions available for our use, we can return to the problem of how to locate the coordinate axes for our model of Alcatraz. Most importantly, we should consider the following question: is there some symmetry in the problem we could use to reduce its complexity?

Although we did not ask this question explicitly, we did consider the question of symmetry in deciding on a conceptual model of the Elysian Fields. In fact, this is how we came to regard flow in the Elysian Fields as 1D. If we were to cut the Elysian Fields on a line perpendicular to the boundaries at the World Encircling

Sea and the River Acheron, we could see that the flow on either side of this cut would be a mirror image of the flow on the opposing side. We could actually make an infinite number of these cuts, and the flow field would be identical in each subdomain to the flow field in any other subdomain, or in the model domain as a whole. It would be wasteful to simulate the entire domain when we can get the same information from a 1D slice, so we chose to represent the Elysian Fields as a 1D problem.

We also considered symmetry when we discussed the possibility of representing Alcatraz as a radial problem. In that case, the symmetry would have been radial, with any 1D radial slice identical to any other slice. Unfortunately, Alcatraz does not possess the correct prerequisites for radial symmetry. However, it may possess other symmetries that we can exploit to reduce the complexity of the problem or the size of the model domain.

Careful examination of Figure 5.3 reveals that the four quadrants of the model domain delineated by the (x_2, y_2) coordinate axes all demonstrate either rotational or reflection symmetry; as a result, we can model any one of the four quadrants and still have the solution for the entire domain. This is an especially important consideration for numerical models, because simulation time is an expense (both in terms of CPU operations and memory requirements), so any steps we can take to cut down on the size of our simulation are welcome. In addition, if we can cut down on the size of our domain using symmetry, we can make a higher resolution grid (if needed) for the same computational cost as a coarser grid of a larger domain that gives us redundant information. You should always seek and use symmetry in your modeling problems.

It is relatively easy to see how the four quadrants of the model domain are symmetric, but we need to consider how we will specify boundary conditions on each of the four sides of any one of the subdomains. Looking at Quadrant I (i.e., $x \geq 0, y \geq 0$), we can see the boundaries at $x = L$ and $y = w$ will be constant head boundaries (Figure 5.4). But what about the boundaries at $x = 0$ and $y = 0$? This is the perfect place for us to apply our boundary conditions of

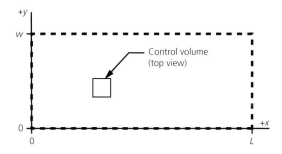

Figure 5.4 Schematic representation of the Alcatraz model domain. The boundaries at $x = L$ and $y = w$ are constant head boundaries (fixed at sea level), while the boundaries at $x = 0$ and $y = 0$ are no-flow (symmetry) boundaries.

the second kind; specifically, no-flow boundaries. We can set these boundaries to no flow because gradients are zero at a symmetry boundary—no flux can cross a symmetry boundary (if it could, flow wouldn't be symmetrical across the boundary). As a result, we can write the boundary conditions for the Quadrant I subdomain as:

$$\frac{dH}{dx}(x = 0, y) = 0, \tag{5.19}$$

$$\frac{dH}{dy}(x, y = 0) = 0, \tag{5.20}$$

$$H(x = L, y) = H_0, \tag{5.21}$$

$$H(x, y = w) = H_0. \tag{5.22}$$

5.4 Quantifying the conceptual model

We have now reached the point where we can begin to quantify our conceptual model. To carry out our plan, we need to apply mass balance to a control volume (CV) located at an arbitrary location within the model domain. Once we have drawn a representative CV, we must label it with all the fluxes going into or coming out of the CV. In this case, we are looking at a 2D problem, so we will have fluxes going into and out of the CV from two different directions; in addition, we have recharge coming through the top surface of our CV. The resulting CV with the fluxes labeled is shown in Figure 5.5.

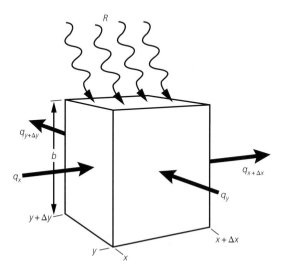

Figure 5.5 Control volume (CV) for the Alcatraz model. Recharge is depicted as coming into the CV from above (R), although mathematically it is generated within the CV. The height of the CV, b [L], is the same as the thickness of the aquifer.

Examination of Figure 5.5 shows two fluxes coming in and two fluxes going out. Note that we always draw our fluxes in the positive axis direction; if fluid is moving in the negative direction, our model will tell us this by giving a negative flux vector. In addition, we have a recharge flux coming through the top side of our CV; we will define recharge coming into the CV as positive by grouping it with the M_{in} terms (negative recharge will then be discharge). Note that, although *conceptually* the recharge is coming in through the top of the CV, *mathematically*, the recharge "appears" inside the CV without transiting any of the CV boundaries. This is a somewhat subtle, but important, distinction. Source/sink terms are sometimes called "generation" terms, because the mass (or energy) associated with them is in some sense generated within the CV; for example, a mass source may be production or injection from a pumping well, or an energy (heat) source may be the result of radioactive decay from a high-level nuclear waste package. The reason this is important is that, in the present (2D horizontal) situation, there would be no distinction drawn between recharge from the surface (e.g., rainfall), upwelling from a deeper aquifer, or discharge from a leaking subsurface pipe— all three situations involve the "generation" of mass within the CV, rather than mass crossing over a boundary of the CV. In terms of units, mass generation is therefore technically specified in units of $[M/L^3T]$, although in this circumstance we can speak rather loosely about recharge having units of $[L/T]$.[4]

Our statement of mass balance is:

$$M_{in} = M_{out} + \Delta M. \qquad (5.23)$$

Expanding each of the terms in Equation 5.23 with reference to Figure 5.5:

$$M_{in} = q_x \Delta y b \rho \Delta t + q_y \Delta x b \rho \Delta t + R \Delta x \Delta y \rho \Delta t, \qquad (5.24)$$

$$M_{out} = q_{x+\Delta x} \Delta y b \rho \Delta t + q_{y+\Delta y} \Delta x b \rho \Delta t, \qquad (5.25)$$

$$\Delta M = 0. \qquad (5.26)$$

In Equation 5.26, we have $\Delta M = 0$ because the problem is defined as steady state. As a result, there will not be any changes in the amount of fluid stored in the CV over time. Checking the units of Equations 5.24 and 5.25, we find the units are [M] (units of mass), which gives us confidence that we have expanded the terms correctly.

Because the quantities ρ (density) and Δt appear in all the terms of Equations 5.24 and 5.25, we can cancel them (dividing through both sides by $\rho \Delta t$). Substituting Equations 5.24 through 5.26 into Equation 5.23 yields:

$$q_x b \Delta y + q_y b \Delta x + R \Delta x \Delta y = q_{x+\Delta x} b \Delta y + q_{y+\Delta y} b \Delta x. \qquad (5.27)$$

Rearranging gives

$$b \Delta y [q_{x+\Delta x} - q_x] + b \Delta x [q_{y+\Delta y} - q_y] = R \Delta x \Delta y, \qquad (5.28)$$

where we have factored the common terms from the fluxes. Dividing through by $b\Delta x\Delta y$, we find

$$\frac{q_{x+\Delta x} - q_x}{\Delta x} + \frac{q_{y+\Delta y} - q_y}{\Delta y} = \frac{R}{b}. \tag{5.29}$$

At this point, we can take $\lim_{\Delta x, \Delta y \to 0}$, and obtain:

$$\frac{\partial q_x}{\partial x} + \frac{\partial q_y}{\partial y} = \frac{R}{b}. \tag{5.30}$$

The next step in our derivation is to substitute Darcy's law for the q terms:

$$\frac{\partial}{\partial x}\left[-K\frac{\partial H}{\partial x}\right] + \frac{\partial}{\partial y}\left[-K\frac{\partial H}{\partial y}\right] = \frac{R}{b}. \tag{5.31}$$

Equation 5.31 is general; that is, it holds for heterogeneous and anisotropic hydraulic conductivity media (although not for variable density fluids). However, we have stated our assumption that the medium is homogeneous and isotropic; as a result, we can remove the hydraulic conductivity from under the differentials to arrive at

$$\frac{\partial^2 H}{\partial x^2} + \frac{\partial^2 H}{\partial y^2} = -\frac{R}{Kb}; \tag{5.32}$$

we can also write this as:

$$\nabla^2 H(x, y) = -\frac{R}{Kb}. \tag{5.33}$$

Note that the recharge term on the right-hand side of Equation 5.32 is divided by the quantity Kb. This is a well-known parameter to most groundwater hydrologists called "transmissivity" [L^2/T] and generally denoted by the symbol T. Transmissivity shows up most often in connection with 2D flow in a horizontal model domain, where it is a convenient expression for the product of hydraulic conductivity and aquifer thickness (as seen here). Although it is sometimes used in other contexts (e.g., flow in fractures), its usefulness is largely restricted to flow in tabular bodies, and it is rarely used in 3D flow situations.

5.5 Nondimensionalization

By this time, the reader is probably familiar with the goals of nondimensionalization, as well as with the approach to arriving at a dimensionless governing equation. However, in nondimensionalizing Poisson's equation, we will see one or two features we have not previously encountered.

When considering how to normalize the dependent variable (head), we first look at the boundary conditions for some indication of the maximum and minimum values head might assume in our problem. In this case, the minimum value is clear (i.e., sea level), but the maximum value (or, equivalently, the range over which head might vary) is not obvious from the boundary conditions.

Because of this uncertainty, we will define a characteristic quantity H_c that will be left arbitrary for the time being. Using this arbitrary characteristic head value and the known minimum head H_0, we can define the dimensionless head (and its derivatives) as:

$$\theta = \frac{H - H_0}{H_c}, \tag{5.34}$$

$$H = H_c\theta + H_0, \tag{5.35}$$

$$dH = H_c d\theta, \tag{5.36}$$

$$d^2H = H_c d^2\theta. \tag{5.37}$$

We can now turn our attention to the independent variables, x and y. In previous chapters, we have only had one independent variable in a given equation—either distance or time. Scanning the ranges of the boundary conditions and the problem geometry shown in Figure 5.4, we can see the domain runs from 0 to L in the x-direction, and from 0 to w in the y-direction. This suggests the following normalization scheme for x:

$$\xi = \frac{x}{L}, \tag{5.38}$$

$$x = L\xi, \tag{5.39}$$

$$x^2 = L^2\xi^2, \tag{5.40}$$

$$dx = Ld\xi, \tag{5.41}$$

$$dx^2 = L^2 d\xi^2, \tag{5.42}$$

and for y it is

$$\zeta = \frac{y}{w}, \tag{5.43}$$

$$y = w\zeta, \tag{5.44}$$

$$y^2 = w^2\zeta^2, \tag{5.45}$$

$$dy = wd\zeta, \tag{5.46}$$

$$dy^2 = w^2 d\zeta^2. \tag{5.47}$$

We can now substitute Equations 5.34 through 5.47 into Equation 5.32:

$$\frac{H_c}{L^2}\frac{\partial^2\theta}{\partial\xi^2} + \frac{H_c}{w^2}\frac{\partial^2\theta}{\partial\zeta^2} = -\frac{R}{Kb}. \tag{5.48}$$

Dividing through by H_c and multiplying by w^2 yields

$$\left(\frac{w^2}{L^2}\right)\frac{\partial^2\theta}{\partial\xi^2} + \frac{\partial^2\theta}{\partial\zeta^2} = -\frac{Rw^2}{KbH_c}. \tag{5.49}$$

We now have the opportunity to define H_c in such a way as to simplify our governing equation by "hiding" a number of dimensional parameters. When faced with a choice of how to define an arbitrary characteristic quantity, it is usually best to choose the definition so as to cancel as many dimensional parameters as possible. In this case, the quantity Rw^2/Kb has dimensions of [L], which are the same dimensions as H_c. As a result, we will set $H_c = Rw^2/Kb$. In addition, the ratio of the width to the length of the domain is already dimensionless; we will therefore define a new dimensionless parameter $\gamma = w/L$. Making the appropriate substitutions, we have

$$\gamma^2 \frac{\partial^2 \theta}{\partial \xi^2} + \frac{\partial^2 \theta}{\partial \zeta^2} = -1. \tag{5.50}$$

We can apply the same variable substitutions to the boundary conditions (Equations 5.19 through 5.22) to find the nondimensional boundary conditions:

$$\frac{\partial \theta}{\partial \xi}(\xi = 0, \zeta) = 0, \tag{5.51}$$

$$\frac{\partial \theta}{\partial \zeta}(\xi, \zeta = 0) = 0, \tag{5.52}$$

$$\theta(\xi = 1, \zeta) = 0, \tag{5.53}$$

$$\theta(\xi, \zeta = 1) = 0. \tag{5.54}$$

Together, the dimensionless Equations 5.50 through 5.54 constitute the governing equation and boundary conditions of our model. An examination of this equation tells us something important about what we can expect from the behavior of our model: in contrast to Laplace's equation, Equation 5.50 is not homogeneous; rather, the derivatives sum to a particular value (in this case, the constant -1). As was stated in Chapter 4, the second derivative can be taken as a function's curvature. The right-hand side of the 2D Equation 5.50, therefore, gives the curvature of θ as a function of ξ and ζ. We can see the curvature is negative because the right-hand side of the equation is negative; as a result, we know the solution plane will be concave-downward (see Figure 5.6). In addition to knowing this from the form of our governing equation, we should have expected curvature in our solution for θ, because our equation has a source term, which is one of the four factors that cause curvature of the potentiometric

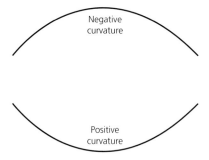

Figure 5.6 Illustration of positive and negative curvature as defined by the second derivative. The upper curve is concave downward, which is negative curvature (as labeled). The lower, concave upward surface shows positive curvature.

surface (Section 4.7.3). We should keep these points in mind when we wish to evaluate our solution(s) to the governing equation.

5.6 Seeking a solution

Now that we have a (dimensionless) governing equation and appropriate boundary conditions, we will, of course, look for a solution to the governing equation that will give us a distribution of θ over the domain $0 \leq \xi \leq 1, 0 \leq \zeta \leq 1$. As it turns out, in this case, arriving at a *complete* analytical solution of Equation 5.50 is more difficult than solving the problems in the previous chapters. We will examine some approaches to finding the full solution of a PDE in future chapters; here, we will content ourselves with using analytical and numerical methods to obtain an *approximate* solution to our problem.

5.6.1 An approximate analytical solution

The dimensionless parameter γ that appeared in Equation 5.50 is a new feature in our process of nondimensionalization. γ is the ratio of the width to the length of the model domain, and it arises as a result of nondimensionalizing a rectangular (as opposed to a square) domain. γ is known as the domain *aspect ratio* and, as is true of all dimensionless parameters, it gives us information about our model system.

By applying a ruler to the scale drawing of the model domain in Figure 5.4, we can find the value of the aspect ratio γ is approximately 5/12, or ~0.4166. Squaring this quantity gives us the coefficient of $\partial^2\theta/\partial\xi^2$ in Equation 5.50, which is $\gamma^2 \approx 0.1736$. What is the significance of this coefficient?

Because of our nondimensionalization of the governing equation, we can be confident that the two terms $\partial^2\theta/\partial\zeta^2$ and -1 in Equation 5.50 are of the order 10^0. However, the term $\gamma^2\partial^2\theta/\partial\xi^2$ is of the order $10^{-1} \times 10^0$, or 10^{-1}. In other words, it is around one order of magnitude less important than the other two terms in the equation. If we can be content with an approximate solution to our problem, we may be able to neglect the $\gamma^2\partial^2\theta/\partial\xi^2$ term in our governing equation. The penalty for doing so will be an error of around 15–20%—roughly the size of the neglected term.

Whether we choose to neglect this term or not is a function of several factors. Primarily, we need to evaluate the level of accuracy our application demands, in light of the likely error that will arise from approximation and the level of effort required to obtain a better solution. Additionally, we may consider where in our solution the errors are likely to manifest themselves. In this case, the errors result from approximating a 2D flow field with a 1D flow field. This will probably be a good approximation near $\xi = 0$, but as $\xi \to 1$, 2D effects will have an increasingly great impact on the flow field. If we primarily need good information near the origin (or near the ζ-axis), a 1D approximation will probably be a good one. Similarly, if our application is relatively insensitive to an error of $\pm 20\%$ over the entire model domain, we can probably live with a 1D estimate. In the case of our protagonist, attempting to work out a groundwater model under pressure on Alcatraz island, it can be convincingly argued that a "quick-and-dirty" estimate is plenty accurate.

If we neglect the first term in Equation 5.50, we have

$$\frac{d^2\theta}{d\zeta^2} = -1, \tag{5.55}$$

which we know we can solve by direct integration, using the same method we used to solve Equation 4.26. Substituting $\chi = d\theta/d\zeta$, the first integration gives

$$\int \frac{d\chi}{d\zeta} d\zeta = \int d\chi = -1 \int d\zeta, \tag{5.56}$$

$$\chi = \frac{d\theta}{d\zeta} = -\zeta + A_1, \tag{5.57}$$

where we have lumped the constants of integration from both sides into A_1. Performing the integration with respect to ζ once again, we find that

$$\int \frac{d\theta}{d\zeta} d\zeta = \int d\theta = \int (-\zeta + A_1)d\zeta, \tag{5.58}$$

$$\theta = -\frac{1}{2}\zeta^2 + A_1\zeta + A_2, \tag{5.59}$$

where we have once again lumped the constants of integration from both sides into one constant; this time, into A_2.

Because we have determined to neglect the $\partial^2\theta/\partial\xi^2$ term in Equation 5.50, we can no longer consider the boundary conditions 5.51 and 5.53. We can, however, apply the boundary conditions 5.52 and 5.54 to find the constants of integration A_1 and A_2. If we take the first derivative of Equation 5.59, we have:

$$\frac{d\theta}{d\zeta} = -\zeta + A_1. \tag{5.60}$$

Applying Equation 5.52, we find that $A_1 = 0$. This makes our solution (Equation 5.59):

$$\theta = -\frac{1}{2}\zeta^2 + A_2. \tag{5.61}$$

Applying the constant head boundary condition (Eq. 5.54), we have $A_2 = +1/2$, so the complete solution for Equation 5.55 is

$$\theta = -\frac{1}{2}\zeta^2 + \frac{1}{2}, \tag{5.62}$$

or

$$\theta(\zeta) = \frac{1}{2}(1 - \zeta^2). \tag{5.63}$$

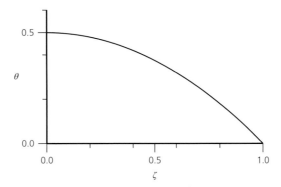

Figure 5.7 Plot of the approximate analytical solution (i.e., 1D flow approximation) for the Alcatraz island problem. $\zeta = 0$ is at the center of the island, while $\zeta = 1$ is the boundary at sea level.

A plot of Equation 5.63 is provided in Figure 5.7. In addition to satisfying the boundary conditions, the model meets our expectations for the curvature of the surface (i.e., $\theta(\zeta)$ has negative curvature). Although this is not a lot of information to use to evaluate our model, at least there are some points that accord with our intuition about how the model should behave. We will return to this solution and its properties in a future section.

5.6.2 A 2D finite difference operator

Although we have an approximate analytical solution, we know it is most likely to apply near the ζ axis; our approximate solution will be progressively less accurate as we move away from $\xi = 0$, and the effects of the boundary condition at $\xi = 1$ begin to assert themselves. We can, of course, seek a more complete analytical solution to our governing equation, and we will take this approach with some problems in future chapters. In this case, however, we will instead look for a way to obtain a more accurate, but still approximate, numerical solution to the governing equation.[5]

In Chapter 4, we developed an expression for the second derivative (Eq. 4.46). We can apply that expression to the two second partial derivatives in Equation 5.50 to arrive at a finite difference version of our governing equation,

$$\gamma^2 \frac{\theta_{i+1,j} - 2\theta_{i,j} + \theta_{i-1,j}}{(\Delta\xi)^2} + \frac{\theta_{i,j+1} - 2\theta_{i,j} + \theta_{i,j-1}}{(\Delta\zeta)^2} = -1, \tag{5.64}$$

where the i subscript indexes node positions on the ξ-axis and j indexes the node positions on the ζ-axis.

5.6.2.1 Iterative solution

In Section 4.8.1, we examined Jacobi and Gauss-Seidel iteration as a means of solving a 1D differential equation. We can apply the same methods to the solution of the 2D finite difference equation. To accomplish this, we must solve

Equation 5.64 for $\theta_{i,j}$. To keep a relatively simple formulation, we will make the assumption that our grid is discretized equally in the ξ and ζ directions; this is not a requirement of the finite difference method, but it does simplify the resulting finite difference operator. Defining $\Delta\xi = \Delta\zeta = \Delta$, we can rearrange Equation 5.64 to obtain:

$$\frac{\gamma^2\theta_{i+1,j} - 2\gamma^2\theta_{i,j} + \gamma^2\theta_{i-1,j} + \theta_{i,j+1} - 2\theta_{i,j} + \theta_{i,j-1}}{\Delta^2} = -1. \tag{5.65}$$

Or, solving for $\theta_{i,j}$,

$$\frac{\gamma^2\theta_{i+1,j} + \gamma^2\theta_{i-1,j} + \theta_{i,j+1} + \theta_{i,j-1}}{\Delta^2} + 1 = \frac{(2\gamma^2 + 2)\theta_{i,j}}{\Delta^2}, \tag{5.66}$$

$$\frac{\gamma^2(\theta_{i+1,j} + \theta_{i-1,j}) + \theta_{i,j+1} + \theta_{i,j-1}}{2(\gamma^2 + 1)} + \frac{\Delta^2}{2(\gamma^2 + 1)} = \theta_{i,j}. \tag{5.67}$$

Note that the $\Delta^2/2(\gamma^2 + 1)$ term on the left-hand side is the source term, which is weighted for the size of the grid cells. For the special case in which there is no source term and the domain is a square (i.e., the aspect ratio $\gamma = 1$), Equation 5.67 reduces to the "classical" 2D finite difference operator:

$$\frac{\theta_{i+1,j} + \theta_{i-1,j} + \theta_{i,j+1} + \theta_{i,j-1}}{4} = \theta_{i,j}. \tag{5.68}$$

The equivalent operator in three dimensions is:

$$\frac{\theta_{i+1,j,k} + \theta_{i-1,j,k} + \theta_{i,j+1,k} + \theta_{i,j-1,k} + \theta_{i,j,k+1} + \theta_{i,j,k-1}}{6} = \theta_{i,j,k}, \tag{5.69}$$

where k is the index for the third spatial axis. The 2D and 3D finite difference operators given in Equations 5.68 and 5.69, respectively, are appropriate for the Laplace equation (i.e., steady-state flow with no source term) with constant and equal grid spacing, and strongly reinforce the Laplacian property of averaging the values of surrounding nodes.

5.7 An alternative nondimensionalization

In Section 5.5, we presented a nondimensionalization scheme that introduced the aspect ratio of the domain into the governing equation. This was quite handy, because it allowed us to determine the conditions under which we could neglect terms relating to one of the spatial dimensions of the problem. Unfortunately, this has the side effect of complicating our finite difference operator (cf. Eqs. 5.67 and 5.68). To avoid this added complexity, we wouldn't typically nondimensionalize in the way demonstrated in Section 5.5 if we intended to find a finite difference solution. Instead, we could normalize both axes on the same characteristic length scale (using the length of the longer axis) to give

$$\xi = \frac{x}{L}, \tag{5.70}$$

$$x = L\xi, \tag{5.71}$$

$$x^2 = L^2\xi^2, \tag{5.72}$$

$$dx = Ld\xi, \tag{5.73}$$

$$dx^2 = L^2 d\xi^2. \tag{5.74}$$

and

$$\zeta = \frac{y}{L}, \tag{5.75}$$

$$y = L\zeta, \tag{5.76}$$

$$y^2 = L^2\zeta^2, \tag{5.77}$$

$$dy = Ld\zeta, \tag{5.78}$$

$$dy^2 = L^2 d\zeta^2. \tag{5.79}$$

The nondimensionalization of H remains the same (see Eqs. 5.34–5.37). Substituting into the dimensional governing equation (Eq. 5.32), we have

$$\frac{H_c}{L^2}\frac{\partial^2\theta}{\partial\xi^2} + \frac{H_c}{L^2}\frac{\partial^2\theta}{\partial\zeta^2} = -\frac{R}{Kb}, \tag{5.80}$$

or

$$\frac{\partial^2\theta}{\partial\xi^2} + \frac{\partial^2\theta}{\partial\zeta^2} = -\frac{RL^2}{KbH_c}. \tag{5.81}$$

We can now set $H_c = RL^2/Kb$, yielding

$$\frac{\partial^2\theta}{\partial\xi^2} + \frac{\partial^2\theta}{\partial\zeta^2} = -1. \tag{5.82}$$

Now, however, when we nondimensionalize the boundary conditions, we find

$$\theta(\xi = 1, \zeta) = \theta(\xi, \zeta = \gamma) = 0, \tag{5.83}$$

$$\frac{\partial\theta}{\partial\xi}(\xi = 0, \zeta) = \frac{\partial\theta}{\partial\zeta}(\xi, \zeta = 0) = 0, \tag{5.84}$$

where $\gamma = w/L$. Given this nondimensionalization (and assuming equal nodal spacing in the ξ and ζ directions), we can find an iterative solution using the simplified finite difference operator:

$$\frac{\theta_{i+1,j} + \theta_{i-1,j} + \theta_{i,j+1} + \theta_{i,j-1}}{4} + \frac{\Delta^2}{4} = \theta_{i,j}. \tag{5.85}$$

When using this formulation, the requirement for equal nodal spacing in both directions dictates careful consideration of the number and spacing of nodes, since the domain is one dimensionless unit long in the ξ-direction, and only γ dimensionless units long in the ζ-direction. If the lengths of the two axes are not commensurable, a more complex formulation will be required that allows different spacings between nodes in the ξ and ζ directions; the derivation of this more general finite difference operator is left for an exercise at the end of this chapter.

5.8 Conclusions

In this chapter, we developed a steady-state model of fluid flow in an area with spatially and temporally constant recharge. The model domain, which nominally represented the island of Alcatraz in the San Francisco Bay, was surrounded by constant potential boundaries; even so, the potentiometric surface exhibited curvature (in this case, convex upward curvature), which is a characteristic of the governing equation—Poisson's equation. We also examined the circumstances under which it may be acceptable to neglect one or more terms in a governing equation; that is, a term may be neglected when it is small (in a nondimensional sense) in comparison with the other terms in the equation, and when the magnitude of the contribution of that term (judged, in a rough sense, by its nondimensional magnitude) is small compared to the accuracy required for the problem's solution.

To assist us with finding the most efficient statement of the problem at hand, we discussed the three main types of boundary conditions (Type 1/Dirichlet, Type 2/Neumann, and Type 3/Robin), and we discussed how the "no-flow" boundary condition (a specialization of Type 2 boundary conditions) can be used to exploit symmetry in the problem. By using existing symmetry, we can often model an entire problem with half (or a quarter, or an eighth, etc.) of the nominal model domain. The use of symmetry saves time and therefore money (the modeler's time, CPU time, memory space, etc.); always seek out and use symmetry in your problems to make the most efficient use of your available resources.

The fact that Poisson's equation contains a source/sink term gave us the opportunity to look at how such a term can be incorporated into a numerical approximation (a finite difference version of the governing equation), which substantially extends the capabilities of our finite difference approximations. In addition, we looked at two possible nondimensionalization schemes for the example problem (the Alcatraz problem): one that used two separate length scales for the independent variables, and was helpful in examining the relative size of the terms in the equation, and a second scheme that used the same length scale to normalize both independent variables. This second scheme is probably more appropriate for most numerical modeling exercises—certainly, it is the one more commonly used in practice.

5.9 Problems

1. Use the approximate analytical solution to the Alcatraz problem (Eq. 5.63) to get an estimate (in nondimensional terms) for the flux of fresh water discharging from the aquifer, out into the bay. What dimensional quantities would you need to know in order to turn this dimensionless flux estimate into a dimensional volumetric discharge?

2. Develop Laplace's equation in two dimensions for the Elysian Fields (Chapter 4) with a finite domain width parallel to the boundaries (i.e., in the direction of the y-axis). Nondimensionalize the equation you derive, and show that a 1D representation of the flow field is appropriate as the width goes to infinity.

3. Write a 2D finite difference operator for $\theta_{i,j}$ for nodal spacing $\Delta\xi \neq \Delta\zeta$. This "variable nodal spacing" finite difference operator is more cumbersome on paper (and in computer code), but it is also more versatile, allowing greater resolution in areas of rapidly changing potential (i.e., steep gradients).

4. On the basis of the "frolicking hypothesis" (see Section 4.3), it seems likely there is little or no rain on the Elysian Fields. On the other hand, if there is grass to frolic on there is certain to be evapotranspiration (ET), and it seems reasonable to hypothesize that the ET increases as one draws closer to the River Acheron (and hence to the Gate of Hell). Assuming the rate of ET from the soil can be described by the equation:

$$f(x) = A_L \sin\left(\frac{\pi x}{2L}\right), \tag{5.86}$$

where L is the distance between the World Encircling Sea and the River Acheron and A_L (units [L/T]) is the amount of ET expected at $x = L$ (i.e., the River Acheron).

   (a) Develop an equation to represent the distribution of head beneath the Elysian Fields while including the effect of ET. HINT: Be careful when defining the directions of your fluxes.

   (b) Solve the governing equation analytically. Compare your solution to the original (Laplace equation) solution. Are the differences significant? Do the differences justify the extra work entailed in the more complex model?

   (c) How does the flux across the Elysian Fields change as a result of incorporating ET into the model?

5. Construct a numerical (finite difference) solution for the 2D Alcatraz problem, and compare your numerical solution with the 1D approximate solution. How refined does your grid have to be (i.e., how small a nodal spacing is required) to get a good match to the approximate analytical solution at $\xi = 0$? In what areas of the model domain is the analytical solution a good approximation, and in what areas is it a poor approximation of the numerical solution?

6. The integral equation

$$A(t)b = Q_i t - S \int_0^t A(t') \frac{dt'}{\sqrt{t - t'}} \qquad (5.87)$$

describes the evolution of wetted area (A) [L^2] in a fracture bounded by an unsaturated matrix. In this equation, A is the dependent variable and t (time) is the independent variable. The remaining quantities are parameters: b is the aperture of the fracture [L], Q_i is the rate at which water is injected to the fracture (considered constant), and S is the "sorptivity" of the matrix (a quantification of the rate at which the matrix takes up water from the fracture) [$L/T^{1/2}$]. Nondimensionalize this equation; if done correctly, the nondimensional form will not have any free parameters. HINT: If you run into difficulties, you can always look up the original paper: Fairley (2010).

Notes

1 An additional problem was the environmental impact of discharging to the San Francisco Bay the raw sewage generated by around 250 prisoners, plus about 60 staff and their families.

2 A well-informed reader would object that the Abbé Faria didn't really dig his way out of the Château d'If with a spoon. In fact, although the Abbé had a number of tools, constructed from everyday objects in his cell, he actually died in prison. Edmond Dantès, on the other hand, did manage to escape—although I will not ruin a superb story by revealing how this was done.

3 Θ is commonly used to represent the fundamental dimension of temperature.

4 Specifying recharge in units of [L/T] is convenient, because precipitation is usually given in, for example, inches/h, cm/h, or similar units. We can get away with this casual usage because, in a horizontal, 2D domain, we multiply the incoming volumetric rate by the density of water and the unit area of the CV to arrive at units of [M/T] being generated in a unit volume.

5 It is sometimes mistakenly stated or believed that numerical solutions are "exact," but this is not so—*every* numerical solution is approximate, although in some cases it may be possible to get arbitrarily good precision from a numerical approximation. This misapprehension is occasionally encountered when reviewers or journal editors demand that an analytical solution to an equation be "verified" by comparison with a numerical solution to the same problem before it can be published.

References

Carslaw, H. and Jaeger, J. (1959) *Conduction of Heat in Solids*, 2nd Edn., Oxford Science Publications, Oxford.

Fairley, J.P. (2010) Fracture/matrix interaction in a fracture of finite extent. *Water Resources Research*, **46**, W08542, doi:10.1029/2009WR008849.

Özişik, M.N. (1993) *Heat Conduction*, 2nd Edn., John Wiley & Sons, Inc., New York.

CHAPTER 6

The transient diffusion equation

Chapter summary

For the past few chapters, we have concentrated on describing the spatial distribution of head within a model domain. Often, however, we are interested in the way the spatial distribution of head evolves over time in response to some perturbation or driver. The Laplace and Poisson equations, being steady-state descriptions of potential distribution, are not adequate for such a task; instead, we must look to a more general description of potential distribution. This more general governing equation is often called the "transient diffusion equation." In the same way that Poisson's equation is a generalization of Laplace's equation (allowing the curvature of the surface to change as a function of position), the transient diffusion equation allows the curvature of the surface to change over time. The diffusion equation is widely used in hydrogeology, heat and mass transport, and many other areas of mathematical physics. While examining some of the properties of this interesting and versatile equation, we will have occasion to learn about one of the oldest methods of solving partial differential equations, and take a brief look at the illustrious and grisly career of a well-known Greek hero.

6.1 The diffusion equation

For the past few chapters, we have contented ourselves with modeling a scalar field (nominally hydraulic head, although the equations apply equally to heat energy, electrical charge, etc.) that is static in time. As we discussed in Chapter 4, Laplace's equation, allows us to calculate the distribution of potential in a field with no sources or sinks and a homogeneous parameter space—hence, the sum of the spatial second derivatives (the surface curvature) is equal to zero. Poisson's equation (Chapter 5), on the other hand, allows for the existence of source/sink terms (and therefore curvature of the solution surface), but the source terms must be steady in time. In order to describe the distribution of potential in a field that changes with time, we must include in our model a temporal derivative that permits the curvature to evolve in response to changing conditions. In this case,

Models and Modeling: An Introduction for Earth and Environmental Scientists, First Edition. Jerry P. Fairley.
© 2017 John Wiley & Sons, Ltd. Published 2017 by John Wiley & Sons, Ltd.
Companion website: www.wiley.com/go/Fairley/Models

the appropriate equation is the "transient diffusion equation," written in three-dimensions (and Cartesian coordinates) as

$$\frac{\partial^2 H}{\partial x^2} + \frac{\partial^2 H}{\partial y^2} + \frac{\partial^2 H}{\partial z^2} = \frac{1}{D}\frac{\partial H}{\partial t}. \tag{6.1}$$

Here D is diffusivity [L^2/T] (the definition of which will be discussed in more detail below). In "del" notation, the diffusion equation is written:

$$\nabla^2 H(x, y, z) = \frac{1}{D}\frac{\partial H}{\partial t}. \tag{6.2}$$

Our examination of the diffusion equation will finally provide a means for us to tie together our investigations of spatially varying potential (in Chapters 4 and 5) and temporally varying quantities (Chapters 2 and 3). At the same time, we will use this opportunity to examine one of the celebrated feats of the Greek half-god, half-man, Hercules.

6.2 The Twelve Labors of Hercules

Hercules (sometimes spelled "Herakles") was one of the Argonauts, and a son of Zeus by a human mother. He lived at a time when the kingdom of Thebes was obliged to pay an unjust tribute to the kingdom of Orchomenos in Boeotia. To get out from under this onerous burden, Hercules led a war against the oppressors of Thebes. Once the war was won, the king of Thebes gave Hercules his daughter, Megara, as his wife. The two lived happily for many years and had three children together. Unfortunately, Hera (both the wife and sister of Zeus; thus, technically, Hercules's step-mother) sent a fit of madness to Hercules, during which he killed his own children.[1] After recovering from his madness, Hercules traveled to the Oracle at Delphi,[2] where the god Apollo ordered him to perform 12 heroic feats, after which he would be forgiven and become immortal. The required feats, which have subsequently become known as the Twelve Labors of Hercules, were to be communicated to him by Eurystheus, king of Tiryns.

6.2.1 The Twelve Labors
There are various accounts detailing the Twelve Labors of Hercules; although they agree on many points, the lists vary somewhat from account to account. Since this is supposed to be a book on modeling rather than a scholarly treatise on Greek mythology, we can make do with the following:

Kill the Nemean Lion. The Nemean Lion was nearly impossible to kill, because arrows couldn't pierce its tough hide. Hercules struck it in the head with his olive-wood club; then, before it could recover, he strangled it with his bare hands. It is said he used the lion's claws to skin it, and ever after wore the skin as a cloak.

Kill the Lernean Hydra. The Hydra was a snake-like monster with many heads; if one of the heads was damaged, two new heads would grow to replace the missing one. Each time Hercules cut off one of its heads, his charioteer and nephew, Iolaus, raced in with a torch and burned the stump before a new head could grow. After this victory, Hercules dipped his arrows in the blood of the Hydra to make them poisonous.

Capture the Cerynian Hind. A hind is a small deer; the Cerynian Hind had golden horns, and was rather a favorite of the goddess Artemis. To avoid incurring the goddess's wrath, Hercules tracked the hind for over a year before he was able to capture it alive. He released it back to the wild, unharmed, after displaying his catch to Eurystheus.

Capture the Erymanthian Boar. The people of Mount Erymanthus lived in fear of this enormous and ferocious pig. Hercules chased the boar into a snowdrift and wrapped it up in a net, after which he carried it back and presented it to the king. Eurystheus was so frightened by the animal that he hid in a giant urn until Hercules took it away (stunts like this did little to endear Hercules to the king).

Clean the Augean Stables. King Augeas owned a very famous herd comprising thousands of cattle. Although the stables had not been cleaned in more than 30 years, Hercules was tasked with cleaning them in a single day. He accomplished this by damming up a nearby river so that the water ran through the stables, washing away the accumulated debris.

Kill the Stymphalian Birds. This flock of obnoxious birds lived (appropriately enough) around Lake Stymphalos. Their beaks were razor sharp, and their feathers fell like arrows, impaling unwary creatures. Hercules used a rattle to frighten them into flight, then killed them with his arrows dipped in the Hydra's blood.

Capture the Cretan Bull. This insanely wild and (reportedly) fire-breathing bull was originally a gift from Poseidon to King Minos of Crete. Hercules captured and dragged it back to Tiryns. With what can only be considered monumentally bad judgment, King Eurystheus set the beast free. It spent the rest of its life terrorizing the Greek population, eventually being killed outside the town of Marathon by Theseus.

Capture the Horses of Diomedes. Diomedes, king of the Bistones, owned a herd of rather ungovernable horses that would only eat human flesh. Hercules and his men made war on the Bistones; when they captured Diomedes, they fed him to his own horses. After that, the horses calmed down, and Hercules made a gift of them to King Eurystheus.

Take the Girdle of the Amazon Queen Hippolyte. Queen Hippolyte originally gave Hercules her beautiful girdle (belt) as a gift for the daughter of Eurystheus. Unfortunately, Hera spread a rumor that Hercules had come to kidnap Hippolyte. The queen was killed in the ensuing mayhem, and Hercules ended up having to fight the entire Amazonian nation to obtain the girdle.

Capture the Cattle of Geryon. Geryon was a monster with wings and three human bodies that owned a beautiful herd of red cattle. The cattle were guarded by a giant and a ferocious two-headed dog. Hercules killed Geryon, the giant, and the dog, and he gave the cattle as a present to Eurystheus.

Take the Golden Apples of the Hesperides. The Hesperides were nymphs (nature spirits) and daughters of the Titan Atlas (one of the progenitors of the Greek gods). The Hesperides owned an orchard in which grew apple trees with golden fruit; the orchard, however, was guarded by a mighty dragon (named Ladon) that had 100 heads. Hercules made a deal with Atlas to hold up the sky while Atlas went to get the apples, but Atlas later refused to take it (the sky) back. Hercules tricked Atlas into taking the sky back "temporarily" while he folded up his cloak (made out of the Nemean Lion) to cushion the weight a little.

Capture Cerberus. As a final labor, Hercules made the arduous trip to the underworld to capture Cerberus, the famed three-headed guard dog of Hades. After some "persuasion," Cerberus agreed to come with Hercules to meet King Erymanthus. Shortly thereafter, Cerberus returned to Hades, none the worse for wear.[3]

6.3 The Augean Stables

Of the 12 labors discussed before, the one that concerns us presently is the cleaning of the Augean Stables. The Augean Stables sat on the floodplain of the Alpheus River (the river that Hercules diverted to wash out the stables). The floodplain was separated from the nearby Peneus River by an area of higher ground, upon which rested King Augeas's palace. A schematic diagram of the area is shown in Figure 6.1.

6.3.1 Developing an approach
As always, we begin to formulate an approach by writing a statement of that which is to be found. In the present circumstance, our interest is primarily to develop an example problem, so we can make do with a simple statement.

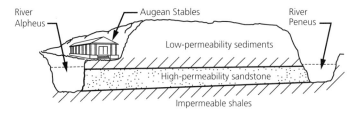

Figure 6.1 Schematic diagram of the Augean Stables problem. The upper and lower layers of low-permeability materials confine a high-permeability aquifer connecting the two rivers.

FIND: The temporal evolution of head in the aquifer between the Alpheus and Peneus rivers, following a perturbation due to an increase in head on the floodplain of the River Alpheus.

If we take the schematic cross section in Figure 6.1 as an indication of what we "know" about the problem, we can quickly make a list of the existing information.

KNOWN:

- The geometry of the problem (e.g., the location and types of boundaries for the domain, the distance between the boundaries, the approximate thickness of the aquifer, etc.)
- The elevations of the rivers (and hence the head at each boundary) and the change in head at time $t = 0$ (i.e., when Hercules dams the River Alpheus)

Most of the assumptions of our model will by now be familiar to you.

ASSUME:

- Homogeneous and isotropic medium.
- Temporally constant medium properties.
- Constant fluid density (we may allow the fluid density to change within the storage term, but not enough to induce density gradient driven flow; see Section 6.4.1).
- Changes in thickness of the aquifer are negligible.
- 1D flow. This looks like a good assumption on the basis of the conceptual cross section (Figure 6.1), but it may not be such a great assumption after we examine the model domain in map view. However, it is a starting point, and we can expand to two or more dimensions as necessary once we understand the 1D results.
- Neglect recharge and discharge. Although recharge or discharge may or may not be important for this problem, it's one additional complexity that we don't need for our first try at modeling the situation. We can always relax this assumption later, when we have a good feel for the expected behavior of the problem.
- Once the head changes at the River Alpheus, it remains constant at its elevated value. It would certainly be possible to construct the model with a time-varying boundary condition, but once again that's an additional factor we can add in later, if necessary.

On the basis of the foregoing, we can sketch out our approach to solving this problem.

APPROACH: Use mass balance to formulate a 1D, transient model of head in the Augean Stables aquifer, subject to an instantaneous change in head at one boundary and a constant head at the other boundary.

6.4 Carrying out the plan

In order to carry out our plan, we first need to figure out where to place our axes; for the present problem, there are no questions of symmetry, and so on, to complicate our choice of grid orientation. Figure 6.2 shows the logical choice for the x-axis, with $x = 0$ at the River Alpheus and $x = L$ at the River Peneus. This is also a good time to adopt a consistent nomenclature for the model (e.g., what symbols will we use for aquifer thickness and water density); in this case, we can stick with the nomenclature we have developed in previous chapters, although we will define any new symbols we require in the course of model development (and there will be a few).

6.4.1 Mass balance and the control volume

To apply mass balance to the problem at hand, we need to define a CV within our model domain—this is a process you should be familiar with by now. Since we have assumed 1D flow, we cannot be expecting any flux in the y- or z-directions, and our aquifer has a well-defined top and bottom. It is therefore appropriate for us to equate the thickness of the aquifer, b [L], with the height of the CV. We don't know the lateral extent of the aquifer (in the y-direction), so the width of the CV (w) will remain arbitrary. In one dimension, there can only be flux in one direction, so we only have one flux arrow going into the CV, and one coming out. As always, we define positive flux to be in the direction of the positive axis, and we have labeled the CV accordingly (Figure 6.3).

As always, our statement of mass balance is given by

$$M_{in} = M_{out} + \Delta M. \tag{6.3}$$

In the steady-state problems we tackled in Chapters 4 and 5, we were able to set the ΔM term to zero because "steady state" implies the amount of material in the CV will not change over any time period. Under that assumption, the mass balance reduces to "mass in equals mass out." Now that we have transient behavior, however, we have opened up the possibility that mass can accumulate in, or disappear from, the CV over time. This phenomenon is more closely related to the change in mass we saw in the Bandurraga Basin problem in Chapter 2.

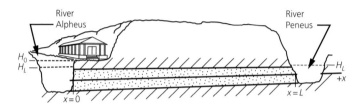

Figure 6.2 Schematic diagram of the Augean Stables problem, showing the x-axis and conditions on the boundaries.

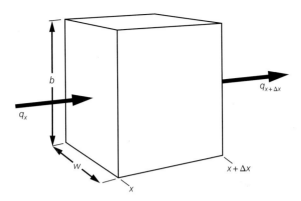

Figure 6.3 Control volume (CV) for the Augean Stables problem. The width of the CV, *w*, is arbitrary, while the height of the CV, *b*, is associated with the thickness of the aquifer.

In fact, it may be useful to think of a CV as a differentially small version of the finite-sized "tank" we defined in the Bandurraga Basin model.

Expanding the three terms in Equation 6.3, we have

$$M_{\text{in}} = q_x w b \rho \Delta t, \tag{6.4}$$

$$M_{\text{out}} = q_{x+\Delta x} w b \rho \Delta t, \tag{6.5}$$

$$\Delta M = \Delta \left[b \rho \phi \right] w \Delta x, \tag{6.6}$$

where ϕ is the porosity [-] of the porous medium (the aquifer). A check on the units of the equations shows all three have units of mass, as desired.

The expansion of this last term (the ΔM term) deserves some explanation. Since the ΔM term describes the change in the mass of stored water, it is within this term that a mechanism must be specified by which mass is lost from or accumulated in the CV. In an unsaturated porous medium, mass of fluid can be added to or subtracted from the CV by changing the saturation of the volume. In this situation, however, we assume we are dealing with a fully saturated medium (which assumption, by the way, is not in our list, but should be added now that we have specified it). As a result, we need a different way to accommodate changes in mass within the CV. The main ways this can be done are (i) through changes in the saturated thickness of the aquifer (*b*); (ii) by changing the porosity of the porous medium (ϕ), either by physically rearranging the grain packing or by allowing an expansion or contraction of the grains themselves; or (iii) by an expansion or contraction of the volume of the water in the pore spaces (i.e., changing the density of the water slightly as a result of changes in stress on the fluid). These three quantities are all included in the square brackets on the right-hand side of Equation 6.6, which brackets are modified by the "Δ" that symbolizes the ability of the included items to change.

In a sense, it would appear that allowing changes in density or the saturated thickness of the aquifer would violate some of our stated model assumptions. However, the essence of the assumptions about saturated thickness and constant density are that the actual changes are small enough that their impact on the flow field is negligible.[4] As a result, we can allow for the tiny changes in density that come with the storage (or release) of elastic energy in water as it is compressed (the compressibility of water is on the order of 4.4×10^{10} m^2/N (Freeze and Cherry, 1979)). Similarly, the compressibility of the aquifer skeleton (ranging from 10^{-6} to 10^{-11} m^2/N, again according to Freeze and Cherry (1979)), or the change in the saturated thickness of the aquifer (usually on the order of millimeters to centimeters), is negligibly small. We can therefore include these parameters in the ΔM term, while excluding them in the remainder of the terms of the mass balance, without introducing appreciable error.

Substituting Equations 6.4–6.6 into Equation 6.3 yields

$$q_x wb\rho \Delta t = q_{x+\Delta x} wb\rho \Delta t + \Delta [b\rho\phi] w\Delta x. \tag{6.7}$$

Dividing through by $wb\rho \Delta x \Delta t$ and rearranging:

$$\frac{q_{x+\Delta x} - q_x}{\Delta x} = -\frac{1}{b\rho} \frac{\Delta [b\rho\phi]}{\Delta t}. \tag{6.8}$$

We can now take $\lim_{\Delta x, \Delta t} \to 0$ to arrive at the differential form of the equation:

$$\frac{\partial q}{\partial x} = -\frac{1}{b\rho} \frac{\partial [b\rho\phi]}{\partial t}. \tag{6.9}$$

By now, the reader should have some idea that the next step is to substitute Darcy's law into Equation 6.9 to get:

$$\frac{\partial}{\partial x}\left[-K\frac{\partial H}{\partial x}\right] = -\frac{1}{b\rho} \frac{\partial [b\rho\phi]}{\partial t}. \tag{6.10}$$

On the basis of our assumption of isotropic and homogeneous materials, we can pull the hydraulic conductivity (K) out of the spatial derivative; rearranging gives us:

$$\frac{\partial^2 H}{\partial x^2} = \frac{1}{Kb\rho} \frac{\partial [b\rho\phi]}{\partial t}. \tag{6.11}$$

Finally, we can apply the chain rule to Equation 6.11 to introduce hydraulic head into the temporal derivative (it's usually best for all the derivatives to refer to the same dependent variable whenever possible),

$$\frac{\partial^2 H}{\partial x^2} = \frac{1}{Kb\rho} \frac{d [b\rho\phi]}{dH} \frac{\partial H}{\partial t}. \tag{6.12}$$

At this point, we need to stop for a brief discussion of storage coefficients—after which we can return to complete the derivation of the transient diffusion equation.

6.4.2 A brief digression on storage coefficients

As we indicated in Section 6.4.1, the mathematical difference between transient and steady-state equations boils down to allowing changes in the quantity of the intensive variable stored in the CV (e.g., water, heat), and the parameter that quantifies this is some type of "storage coefficient." In heat transfer, the storage coefficient is the specific heat; in chemical transport, the storage coefficient is not usually broken out separately, but is part of the medium diffusivity. In hydrogeology, there are a wide variety of storage coefficients, each slightly different from the others and developed to apply in a specific situation. The most commonly encountered of these coefficients are the "storativity," "specific storage," and "specific yield."

Going back to Equation 2.8, it is easy to imagine that the term $1/b\rho\ d[b\rho\phi]/dH$ in Equation 6.12 is a type of storage coefficient. This term is a quantification of how much water is released from storage due to a change in head, and therefore due to a change in the effective stress in the aquifer. This change in the effective stress changes the quantities b, ρ, and ϕ, resulting in an expulsion or accumulation of water. It is easier to see how this works if we use the product rule to expand the derivative:

$$\frac{1}{b\rho}\frac{d[b\rho\phi]}{dH} = \frac{\phi}{b}\frac{db}{dH} + \frac{\phi}{\rho}\frac{d\rho}{dH} + \frac{d\phi}{dH}, \tag{6.13}$$

$$\frac{1}{b\rho}\frac{d[b\rho\phi]}{dH} = \phi\left[\frac{1}{b}\frac{db}{dH} + \frac{1}{\rho}\frac{d\rho}{dH} + \frac{1}{\phi}\frac{d\phi}{dH}\right]. \tag{6.14}$$

You should try this yourself to verify the correctness of the expansion.

Examination of Equation 6.14 reveals that each term is tied to a physical process ($b \rightarrow$ aquifer compaction; $\rho \rightarrow$ volumetric expansion of the water; $\phi \rightarrow$ change in the volume of pore space). The changes are normalized on their original values such that the amount of water released from storage is appropriate for a unit volume. The derivatives are also multiplied by the volume of pore space (ϕ), because the water is only released from that fraction of the volume occupied by fluid. The sum of these terms, therefore, can be said to quantify "the volume of water released from a unit volume of aquifer in response to a unit change in head," which is the definition of "specific storage" (S_s). From Equation 6.14, it should be clear that the units of specific storage are [1/L].

Specific storage is the most versatile of the storage terms used for fully saturated media. Because it is defined on a *per unit volume* basis, it is applicable to problems in one, two, or three dimensions. On the other hand, storativity (S) is applicable only to 1D or 2D problems, where the aquifer is horizontally extensive and vertical gradients can be neglected. The reason for this is that storativity is defined as the amount of water released from storage by a unit drop in head *per unit area of aquifer*. Storativity is related to specific storage through the simple relationship:

$$S = S_s b, \tag{6.15}$$

where b is the saturated thickness of the aquifer. Storativity is unitless [-], and values of storativity are generally found to be in the range of 5×10^{-3} to 5×10^{-5} (Freeze and Cherry, 1979).

Although storativity is limited to use in horizontal model domains, it is nonetheless commonly used by professional hydrogeologists, and the value of storativity for a particular aquifer or aquifer test is much more likely to be reported than the value of specific storage. There are a number of reasons for this; perhaps most obvious is that hydrogeologists tend to conceptualize aquifers as horizontal, tabular bodies, and storativity is the appropriate parameter for such formations. A quick check will show that the problems investigated in this text tend to fall into this conceptualization. Other reasons for using storativity are that most aquifer pump tests are parameterized in terms of storativity and transmissivity, T, $[L^2/T]$, where transmissivity is equal to

$$T = Kb. \tag{6.16}$$

Transmissivity is to hydraulic conductivity as storativity is to specific storage; that is, it is a 1D or 2D representation that is applicable to a conceptually horizontal, tabular aquifer. For this reason, storativity and transmissivity are usually used together.

Another reason storativity (and transmissivity) are often favored by hydrogeologists is that the actual saturated thickness of an aquifer is rarely well constrained. By lumping the aquifer thickness together with the specific storage (and/or hydraulic conductivity), we bundle several sources of uncertainty into one effective parameter, similar to the effective parameters we defined for the Bandurraga Basin problem (Chapter 2). Storativity and transmissivity are common and useful parameters, and you should be familiar with their uses while being aware of their limitations. However, it is worth noting that, although 2D tabular representations of aquifers are ubiquitous and analytically tractable, not all aquifers can be reasonably described with this simple geometry. Always be on the lookout for potential situations that do not fit this mold, and don't fall into the trap of assuming every aquifer you model has what is euphemistically known as "layer cake" hydrology.

The third term mentioned in the opening paragraph of this section, specific yield (S_y), is related to storativity, but it applies to water table (unconfined) aquifers. Specific yield is the quantity of water produced from storage from a unit area of aquifer (i.e., vertically integrated over the entire thickness b of the aquifer) per unit drop in head. However, because specific yield applies to unconfined aquifers, the water produced from a drop in head is primarily derived from drainage of pore space. Because actual draining of the pores produces a much greater volume of water than would be realized from a commensurate drop in head in a confined aquifer, values of specific yield are usually orders of magnitude greater than the values of storativity associated with confined aquifers. Values of specific yield commonly range from 0.01 to 0.3 (Hermance, 1999). Because specific yield is so much greater than storativity, the cone of depression from a pumping well (see Chapter 7) in a confined aquifer is almost always of much greater extent for a given rate of pumping than a similar well, pumping at a similar rate for the same length of time, in an unconfined aquifer.

Occasionally, the phrase "storativity of an unconfined aquifer" is used. This phrase refers to the arithmetic sum of specific yield and the specific storage of the aquifer times the saturated thickness (Hermance, 1999).

$$S = S_y + bS_s; \tag{6.17}$$

however, unless the specific storage is large and the saturated thickness of the aquifer is much greater than the quantity of water released from storage due to pore drainage, the second term in Equation 6.17 is usually negligible in comparison to the first term. Similar to (confined) storativity, specific yield and unconfined storativity are both dimensionless quantities.

Before returning to complete the derivation of the governing equation for the Augean Stables problem, it is worthwhile to spend a few minutes of thought on the fact that all the storage coefficients discussed in this section (and all storage coefficients commonly in use) are assumed to be equal to a constant. This assumption is rarely discussed or even given serious consideration by most modelers; however, a moment's thought will show that storage coefficients are probably only equal to constants in very limited circumstances. That is, the assumption of a storage coefficient being equal to a constant is likely to hold for small changes in head, but is questionable for larger changes in head. The most obvious example of this is the pumping of a confined aquifer until the potentiometric surface is drawn down below the top of the aquifer, resulting in a "semi-confined" or "partially confined" situation. Once the potentiometric surface decreases below the confining layer, actual pore drainage begins, and the appropriate storage coefficient for modeling the aquifer would then be a combination of storativity (in those areas of the aquifer that were still confined) and specific yield (in the unconfined portions). Furthermore, the storage coefficient will continue to change as long as the unconfined area of the aquifer continues to expand.

Other situations where the storage coefficient of an aquifer is not a constant can also be imagined. The most important of these situations occurs in aquifers that comprise unlithified sediments. Elastic changes in the aquifer thickness or changes in the density of fluid in the pore space due to changes in the effective stress are reversible; however, the linear behavior implied by a constant storage coefficient can only hold over a very limited range of change in head. Changes in the pore structure of unconsolidated sediments may be responsible for a portion of the observed value of the storage coefficient and, although elastic deformation is reversible, physical grain rearrangement and the resulting reduction in porosity are invariably irreversible. Thus, subsequent tests may show a reduction in the storage coefficient over the original test.

The assumption of constant storage coefficients is common and useful. Problems using constant storage expressions are analytically more tractable than the alternative—variable coefficients certainly make the problem more difficult, and may cause the governing equation to be nonlinear. Despite the fact that you will make use of this assumption in virtually all saturated zone models you develop, you should never lose sight of the fact that *this is an assumption*. As with all assumptions, you should be aware of the implications, and you should be alert to indications that it may not hold true for a given situation.

6.4.3 Completing the governing equation

After considering carefully the various types of storage coefficients available to us (Section 6.4.2), it should be clear that storativity is the correct coefficient to use for the Augean Stables problem. The storage term in Equation 6.12 is specific storage, however. To change over to storativity, we can multiply the left-hand side of Equation 6.12 by b/b to obtain

$$\frac{\partial^2 H}{\partial x^2} = \frac{S_s b}{Kb} \frac{\partial H}{\partial t}. \tag{6.18}$$

Comparing Equation 6.18 with Equations 6.15 and 6.16, we see we can rewrite the equation as

$$\frac{\partial^2 H}{\partial x^2} = \frac{S}{T} \frac{\partial H}{\partial t}, \tag{6.19}$$

where S is storativity and T is transmissivity. Alternatively, the combination T/S is known as the "hydraulic diffusivity," D [L^2/T] (or often simply called "the diffusivity"), and the dimensional governing equation can be written as:

$$\frac{\partial^2 H}{\partial x^2} = \frac{1}{D} \frac{\partial H}{\partial t}. \tag{6.20}$$

The version of the governing equation shown in Equation 6.19 is the form most commonly seen in hydrology and hydrologic modeling, but the second version (Equation 6.20) is usually preferred in heat and mass transport applications. The astute modeler should be familiar with both styles, since published solutions to a problem of interest may be found in either form.

Although we have a dimensional governing equation, the reader should be aware by now that, in order to find a complete solution to the problem (i.e., a solution specific to the problem at hand), we will require some auxiliary conditions. As was stated earlier, the general rule is that a partial differential equation needs a condition for each order of each derivative with respect to each independent variable. In this case we have a second-order derivative in space and a first-order derivative in time, implying the need for two spatial conditions and one temporal condition—in other words, two boundary conditions and one initial condition. Thinking about the problem statement and looking at Figure 6.2, we can state the appropriate conditions as:

$$H(x, t < 0) = H_L; \tag{6.21}$$

$$H(x = 0, t) = H_0; \tag{6.22}$$

$$H(x = L, t) = H_L. \tag{6.23}$$

where H_L indicates the level of head at the Peneus River (at $x = L$) and across the entire model domain at all times $t < 0$, and H_0 is associated with the assumed instantaneous change in water level (head) at the Alpheus River for all times $t \geq 0$. Together, the governing Equation (Eq. 6.19 or 6.20) and the auxiliary conditions given in Equations 6.21 through 6.23 define a well-posed statement of our model of the Augean Stables aquifer.

6.4.4 Nondimensionalization

We have now come to the point in our investigation at which we can begin to nondimensionalize the governing equation and boundary conditions. Following our usual strategy, we first look to the auxiliary conditions for an indication of how to normalize the dependent variable H. From examination of Equations 6.21 through 6.23, we find there are clearly defined minimum and maximum values for head—it should vary between H_L (the likely minimum value) up to H_0 (the likely maximum value). As a result, we can use the by now familiar normalization scheme:

$$\theta = \frac{H(x, t) - H_L}{H_0 - H_L}. \tag{6.24}$$

To change variables in the governing equation, we will need to solve Equation 6.24 for θ, as well as for the first and second derivatives of θ:

$$H(x, t) = (H_0 - H_L)\theta + H_L, \tag{6.25}$$

$$dH = (H_0 - H_L)d\theta, \tag{6.26}$$

$$d^2H = (H_0 - H_L)d^2\theta. \tag{6.27}$$

There is also an obvious length scale in the boundary conditions on which to normalize the spatial (independent) variable. We therefore define the nondimensional spatial axis as:

$$\xi = \frac{x}{L}. \tag{6.28}$$

Solving for x and its derivatives yields

$$x = L\xi, \tag{6.29}$$

$$x^2 = L^2\xi^2, \tag{6.30}$$

$$dx = Ld\xi, \tag{6.31}$$

$$dx^2 = L^2d\xi^2. \tag{6.32}$$

Finally, we come to the temporal independent variable. In our problem, time runs from $0 \leq t < \infty$, which, as we have seen in previous examples, offers no clear characteristic time on which to normalize. Following our nondimensional heuristic (Appendix A), we define an arbitrary characteristic time t_c and normalize on this to-be-determined value:

$$\tau = \frac{t}{t_c}, \qquad (6.33)$$

$$t = t_c \tau, \qquad (6.34)$$

$$dt = t_c d\tau. \qquad (6.35)$$

We are now in a position to substitute into our dimensional governing equation; in doing so, we find

$$\frac{(H_0 - H_L)}{L^2} \frac{\partial^2 \theta}{\partial \xi^2} = \frac{(H_0 - H_L)S}{Tt_c} \frac{\partial \theta}{\partial \tau}. \qquad (6.36)$$

Canceling the $(H_0 - H_L)$ terms and multiplying through by L^2 leaves

$$\frac{\partial^2 \theta}{\partial \xi^2} = \frac{L^2 S}{Tt_c} \frac{\partial \theta}{\partial \tau}. \qquad (6.37)$$

Examining the units of $L^2 S/T$ shows they are equal to $[T]$; as a result, we can set the characteristic time equal to

$$t_c = \frac{L^2 S}{T} = \frac{L^2}{D}, \qquad (6.38)$$

which gives a final, dimensionless form of the governing equation

$$\frac{\partial^2 \theta}{\partial \xi^2} = \frac{\partial \theta}{\partial \tau}. \qquad (6.39)$$

A similar substitution on the boundary and initial conditions gives

$$\theta(\xi, \tau = 0) = 0, \qquad (6.40)$$

$$\theta(\xi = 0, \tau) = 1, \qquad (6.41)$$

$$\theta(\xi = 1, \tau) = 0. \qquad (6.42)$$

Together with the definitions of the nondimensional variables, Equations 6.39 through 6.42 constitute the dimensionless statement of the governing equation of our problem.

6.5 An analytical solution

In this section, we will develop an analytical solution for Equation 6.39, subject to the conditions specified in Equations 6.40 through 6.42. The method we will use is known as "separation of variables," and it is one of the oldest methods used to obtain a solution of a PDE. Separation of variables does not work on all PDEs; however, it is a powerful technique that does work with many PDEs, and it deserves a place in most modelers' bag of tricks. On the other hand, I recognize that not every modeler has a predilection or the temperament for developing analytical solutions to PDEs. For those that have little interest in the details of the mathematics, I recommend reading Section 6.5.1 to gain an overall understanding of the method. The reader may then skip to Section 6.6 to see the way in which the solution is implemented, and gain some insight into the physical meanings of the various parts of the solution.

For those readers that have the interest and patience for the somewhat detailed mathematics, the method of separation of variables has much to offer toward a greater understanding of PDEs, and mathematical models in general. In particular, the key technique of the method, which is the construction of a Fourier series of the solution using the property of orthogonality, is a powerful and pervasive technique in mathematical physics. A little study of the methods illustrated here will pay dividends in your ability to understand the work of others and communicate with them as equals.

6.5.1 Separation of variables: The basic idea

As we mentioned in Section 6.5, separation of variables is one of the oldest and best-known methods for solving PDEs, having first been used by L'Hopital in 1750. The essence of the separation of variables method is to assume that the dependent variable can be represented as a multiplicative combination of independent functions, each of which is a function of one independent variable. For example, in the case of a function $f(x, y, z)$, we assume there exist three functions, Ξ, Υ, and Ω, such that

$$f(x, y, z) = \Xi(x)\Upsilon(y)\Omega(z). \tag{6.43}$$

This may or may not be the case; however, if it turns out to be true for the PDE we are interested in, it is likely we can decompose the original PDE into three ordinary differential equations (ODEs). ODEs are usually easier to solve than PDEs; if we can solve the three ODEs (and find a way to match the boundary conditions), we can multiply the solutions of the three ODEs together to obtain the solution to the original PDE.

Of course, there are lots of ins and outs to the separation of variables process, and even after the method has been mastered, not all PDEs are amenable to this type of solution. It sometimes happens that the governing equation is separable, but the boundary conditions are not. Also, if the governing equation is non-homogeneous (i.e., if the equation includes a source/sink term), the separation process is generally much more involved (or may not, in some circumstances, be possible at all). However, the general method is applicable to a wide range of PDEs; furthermore, it doesn't require tedious or difficult back-transformations using transform tables or contour integration, as do the more powerful transform methods (e.g., Laplace transforms and Fourier transforms). As a result, it is worthwhile for an aspiring modeler to familiarize her/himself with the separation of variables method. In Sections 6.5.2 and 6.5.3, we present one relatively simple example of the technique, along with (hopefully) instructive notes on the process; for more insight into the method and its extension to heterogeneous equations, as well as an introduction to the transform methods and other techniques for solving PDEs, I highly recommend the understandable and clearly written presentation in Farlow (1993). Another excellent book on the solution of PDEs is that of Logan (2004); for an introduction to the Laplace transform and its application to ODEs, I recommend Logan (2006).

6.5.2 Initial preparations

Before starting on the separation of variables proper, we need to do a little preparation of the governing equation and boundary conditions to get them in a format for which separation of variables will work. Specifically, separation of variables will only work if all but one of the auxiliary conditions are homogeneous (which is the case with our problem); in addition, we would like the nonhomogenous condition to be the initial condition[5] (which is not the case for our problem). Getting our equation and boundary conditions into the proper configuration is therefore something we must achieve before beginning the separation process.

To get our equation and conditions into the desired format, we can think of our solution as consisting of two parts: a transient part and a steady-state part. Farlow (1993) provides a general approach to this decomposition, using the function

$$\theta = A\xi + B(1 - \xi), \tag{6.44}$$

as the steady portion of the solution, and finding the constants A and B to satisfy the boundary conditions on ξ. However, in this case, we can quickly see the "final" (steady) configuration of the potentiometric surface will be linear, running from a maximum of $\theta = 1$ at $\xi = 0$ to a minimum of $\theta = 0$ at $\xi = 1$. This follows because the temporal derivative $\partial\theta/\partial\tau$ must be equal to zero at steady state. Setting the temporal derivative equal to zero gives us the same governing equation as the Elysian Fields problem (see Equation 4.26), and we know the solution to that problem, given by Equation 4.37, is

$$\theta = 1 - \xi, \tag{6.45}$$

which accords with our physical intuition for the late-time behavior of the Augean Stables aquifer. Furthermore, any function that satisfies the governing equation is a necessary part of the complete solution, and Equation 6.45 does satisfy the governing equation (Eq. 6.39), as you should verify. We can therefore split the complete solution for θ into two portions: a steady-state part (θ_s) and a transient part (θ_t):

$$\theta = \theta_s(\xi) + \theta_t(\xi, \tau) = 1 - \xi + \theta_t(\xi, \tau). \tag{6.46}$$

Substituting into the governing equation and performing the indicated operations yields

$$\frac{\partial^2}{\partial \xi^2}[1 - \xi + \theta_t] = \frac{\partial}{\partial \tau}[1 - \xi + \theta_t],$$

(6.47)

$$\frac{\partial^2 \theta_t}{\partial \xi^2} = \frac{\partial \theta_t}{\partial \tau}.$$

(6.48)

A similar substitution on the initial and boundary conditions gives:

$$\theta_t(\xi, \tau = 0) = \xi - 1,$$

(6.49)

$$\theta_t(\xi = 0, \tau) = 0,$$

(6.50)

$$\theta_t(\xi = 1, \tau) = 0.$$

(6.51)

You should verify for yourself that the transformed governing equation and auxiliary conditions are correct.

6.5.3 Separation of variables: The method

Using the transformed governing equation and auxiliary conditions as a starting point, we begin the separation process by assuming the dependent variable, θ_t, can be decomposed into two functions we will call Ξ and Γ, each of which is a function of only one independent variable,

$$\theta_t(\xi, \tau) = \Xi(\xi)\Gamma(\tau).$$

(6.52)

If our hypothesis is true, we can write the derivatives in the governing equation (Eq. 6.39) as:

$$\frac{\partial^2 \theta_t}{\partial \xi^2} = \frac{\partial}{\partial \xi^2}[\Xi(\xi)\Gamma(\tau)] = \Xi''\Gamma,$$

(6.53)

$$\frac{\partial \theta_t}{\partial \tau} = \frac{\partial}{\partial \tau}[\Xi(\xi)\Gamma(\tau)] = \Xi\Gamma'.$$

(6.54)

where the primes indicate differentiation with respect to the appropriate independent variable. Substituting Equations 6.53 and 6.54 into the governing equation yields:

$$\Xi''\Gamma = \Xi\Gamma'.$$

(6.55)

If we divide both sides of Equation 6.55 by $\Xi\Gamma$, we have:

$$\frac{\Xi''}{\Xi} = \frac{\Gamma'}{\Gamma}.$$

(6.56)

This is a somewhat odd statement, since it must be true regardless of what values of ξ and τ are substituted into the functions Ξ and Γ, respectively. It turns out that the only way this can be so is if both functions are equal to the same constant,

$$\frac{\Xi''}{\Xi} = \mu = \frac{\Gamma'}{\Gamma},$$

(6.57)

where μ is called the *separation constant*. At this point, the value of μ is arbitrary; that is, it may be positive, negative, or equal to zero. To make this clear, it is customary to define a new, squared[6] separation constant that is explicitly positive or negative (or zero) and equal to μ as:

$$\pm\lambda^2 = \mu. \tag{6.58}$$

Equation 6.57 is therefore identical with

$$\frac{\Xi''}{\Xi} = \pm\lambda^2 = \frac{\Gamma'}{\Gamma}. \tag{6.59}$$

Since both sides of Equation 6.59 are a function of one independent variable only, with some rearranging we can write the two sides as individual equations:

$$\Xi'' \pm \lambda^2 \Xi = 0, \tag{6.60}$$
$$\Gamma' \pm \lambda^2 \Gamma = 0. \tag{6.61}$$

Equation 6.60 requires two boundary conditions; we can write this system as:

$$\Xi'' \pm \lambda^2 \Xi = 0, \tag{6.62}$$
$$\Xi(\xi = 0) = 0, \tag{6.63}$$
$$\Xi(\xi = 1) = 0. \tag{6.64}$$

We will first concentrate on solving Equation 6.62, and then return to Equation 6.61 and the remaining (initial) condition.

6.5.4 Solving for spatial dependence

To solve Equation 6.62, we must consider three possible cases: $\lambda^2 = 0$, $\lambda^2 < 0$, and $\lambda^2 > 0$. Some of these cases will be ruled out because they cannot match the boundary conditions; the remaining cases that do satisfy the boundary conditions will be added together to construct the full function Ξ.

6.5.4.1 Case 1: $\lambda^2 = 0$

When $\lambda^2 = 0$, Equation 6.62 reduces to

$$\Xi'' = 0, \tag{6.65}$$

which has as its solution:

$$\Xi = A\xi + B. \tag{6.66}$$

The only combination of constants A and B that can reproduce the boundary conditions (Eqs. 6.63 and 6.64) is $A = B = 0$, which is known as "the trivial case." As a result, this solution cannot be part of the complete solution of Equation 6.62.[7]

6.5.4.2 Case 2: $\lambda^2 < 0$

In the case where $\lambda^2 < 0$, the equation for Ξ becomes

$$\Xi'' - \lambda^2 \Xi = 0. \tag{6.67}$$

Solutions to this equation have the form

$$\Xi = A \sinh(\lambda\xi) + B \cosh(\lambda\xi), \tag{6.68}$$

as can be verified by differentiating twice and substituting into Equation 6.65. Again, however, when we apply the boundary conditions (Eqs. 6.63 and 6.64), it is clear that the only values of A and B that can satisfy the boundary conditions are $A = B = 0$. Once again, this is the trivial case, and will not contribute to the solution of our equation.

6.5.4.3 Case 3: $\lambda^2 > 0$

In the case where $\lambda^2 > 0$, the equation for Ξ becomes

$$\Xi'' + \lambda^2 \Xi = 0. \tag{6.69}$$

Solutions to this equation have the form

$$\Xi = A \sin(\lambda\xi) + B \cos(\lambda\xi), \tag{6.70}$$

which can again be verified by differentiating twice and substituting into Equation 6.65. In this case, however, we finally find the possibility of meeting the boundary conditions and thus arriving at a complete solution of our problem. Applying the first condition (Eq. 6.63), we have

$$A \sin(\lambda 0) + B \cos(\lambda 0) = 0; \tag{6.71}$$

from which it is clear that $B = 0$. Next, applying Equation 6.64, we find that

$$A \sin(\lambda) = 0. \tag{6.72}$$

Of course, we could set $A = 0$ to make this equation true; however, that would result once again in the trivial case, and we need to avoid that in order to have any solution other than $\Xi = 0$. In this situation, however, we have another possibility: if the constant λ, which is still arbitrary, is equal to any of the values,

$$\lambda = \pi, 2\pi, 3\pi, \ldots, \tag{6.73}$$

then Equation 6.72 will meet the boundary conditions. In fact, because all integer multiples of π are allowable solutions, we will eventually have to sum them all to achieve the complete solution, but for now we will be content with a statement of all the solutions,

$$\Xi_n(\xi) = A_n \sin(n\pi\xi), \quad n = 1, 2, 3, \ldots. \tag{6.74}$$

Now that we have the solution to the spatial portion of the governing equation, we can move on to the temporal equation. We will come back to Equation 6.74 to deal with the constants A_n later.

6.5.5 Solving for temporal dependence

Equation 6.61 gives the ordinary differential equation describing the temporal dependence of the Augean Stables governing equation. In fact, we already know, based on our work with the spatial part of the equation (shown in Section 6.5.4), that $\lambda^2 > 0$, so we can rewrite Equation 6.61 as

$$\Gamma' + \lambda^2 \Gamma = 0; \tag{6.75}$$

or, to put this equation in a more recognizable form, we can rearrange slightly (and change notation) to find

$$\frac{d\Gamma}{d\tau} = -(n\pi)^2 \Gamma, \tag{6.76}$$

This equation should look very familiar to you, because it is essentially the equation for exponential decay (Eq. 2.20) from Section 2.4. Following the solution method outlined in Section 2.5, we find that:

$$\Gamma_n(\tau) = C_n e^{-(n\pi)^2 \tau}, \quad n = 1, 2, 3, \ldots, \tag{6.77}$$

where the "C_n" are constants of integration. As with Equation 6.74, we will content ourselves for the time being with listing all the solutions, and leave their summation for a later section.

6.5.6 Summing up

Following from Equation 6.52, we know that any individual solution for θ_t is found by multiplying the individual solutions for Ξ and Γ. As a result, we can write

$$\theta_{t,n} = D_n e^{-(n\pi)^2 \tau} \sin(n\pi\xi), \quad n = 1, 2, 3, \ldots, \tag{6.78}$$

where $D_n = A_n C_n$.

As was mentioned earlier, however, in order to find a complete solution to a differential equation, it is necessary to add together all possible solutions that fit the auxiliary conditions. We know that Equation 6.78 fits the boundary conditions for all values of n because we purposely constructed it that way (although our solution hasn't yet been checked against the initial condition, we will do that in a section or two). Our solution, therefore, comprises an infinite series given by

$$\theta_t = D_1 e^{-\pi^2 \tau} \sin(\pi\xi) + D_2 e^{-(2\pi)^2 \tau} \sin(2\pi\xi) + \ldots$$
$$+ D_n e^{-(n\pi)^2 \tau} \sin(n\pi\xi) + \cdots . \tag{6.79}$$

There are, of course, two final obstacles that must be overcome to complete the solution to our problem: we must find a way to make Equation 6.79 match the initial condition, and we must assign values to an infinite number of constants (the D_n constants). It turns out that these two problems are related, and we will be in an excellent position to attack them after a brief discussion of Fourier series.

6.5.7 Joseph Fourier and the Augean Stables

You may be wondering what Monsieur Joseph Fourier has to do with the Augean Stables; in fact, he was the first person to solve the Augean Stables problem (although, of course, he didn't think of it as "the Augean Stables problem"). According to Bell (1937), Fourier was at Grenoble, France, when he composed his work *Theorie analytique de la chaleur* (*The Mathematical Theory of Heat*) in 1807. The French Academy was so impressed with his work that, in 1812, they set as the Grand Prize problem the development of a theory that could model the change of temperature in a thin (i.e., 1D) metal rod, initially at a given temperature and subject to a sudden change in temperature at one boundary. Although Fourier was working with heat energy diffusion, the problem is mathematically identical to the Augean Stables problem in hydrology. Fourier did win the French Grand Prize for his heat transfer work; unfortunately, the lack of mathematical rigor in his theory set him up for serious criticism from the referees (either Laplace, Lagrange, and Legendre according to Bell (1937), or Laplace, Lagrange, Monge, and LaCroix, according to Lathi (1998)). Ultimately (over 100 years later (Bell, 1937)), pure mathematicians finally managed to clear away the theoretical objections to the use of trigonometric series (commonly referred to as "Fourier series"); Fourier series are now the cornerstone of mathematical analysis.[8]

The idea behind Fourier series is that (almost) any periodic function can be represented by a series of sines and cosines:

$$f(x) = a_0 + a_1 \sin\left(\frac{\pi x}{p}\right) + b_1 \cos\left(\frac{\pi x}{p}\right) + \cdots$$
$$+ a_n \sin\left(\frac{n\pi x}{p}\right) + b_n \cos\left(\frac{n\pi x}{p}\right) + \cdots . \tag{6.80}$$

In 1753, Daniel Bernoulli proposed essentially the same series as a general solution to the wave equation, but he unfortunately offered no method for determining the coefficients of the series (Stillwell, 2002). The method for finding the coefficients of the series is not trivial, but once Fourier presented the basic method, the floodgates were opened to apply this versatile tool in a wide range of pure and applied mathematics, including mathematical physics, signal processing, number theory, and Cantor's theory of sets (Stillwell, 2002).

6.5.8 Orthogonality

The key to determining an infinite number of coefficients that will make a trigonometric series equal to an arbitrary periodic function is a property, possessed by some families of functions, known as "orthogonality." A sequence of real functions f_n, where $n = 1, 2, 3, \ldots$, that are defined on an interval (a, b), is said to possess the property of orthogonality if they satisfy the equation (Wylie and Barrett, 1995):

$$\int_a^b f_n(x) f_m(x)\, dx = \begin{cases} = 0 & m \neq n \\ \neq 0 & m = n \end{cases} \tag{6.81}$$

Although Equation 6.81 may not look like a very clear definition of anything, it is crucial to the calculation of Fourier coefficients, as will be shown later. The property of orthogonality, however, is much more broadly applicable than simply calculating Fourier coefficients. Lots of functions possess the property of orthogonality; for example, Bessel functions, Hermite and Legendre polynomials, and spherical harmonics. We won't discuss any orthogonal functions other than the trigonometric series (i.e., Fourier series), or get into a deep discussion of orthogonality and its applications in other areas. In terms of this problem, however, the orthogonality of trigonometric series will play an important role in allowing us to complete our solution of the Augean Stables model.

To see how orthogonality can be used to find the coefficients of a Fourier series, we begin by applying the initial condition to Equation 6.79 to find

$$\xi - 1 = D_1 e^{-\pi^2 0} \sin(\pi \xi) + D_2 e^{-(2\pi)^2 0} \sin(2\pi \xi) + \cdots$$
$$+ D_n e^{-(n\pi)^2 0} \sin(n\pi \xi) + \cdots, \tag{6.82}$$

which reduces to

$$\xi - 1 = D_1 \sin(\pi \xi) + D_2 \sin(2\pi \xi) + \cdots + D_n \sin(n\pi \xi) + \cdots. \tag{6.83}$$

Since the orthogonal functions we wish to exploit are sine functions, we multiply both sides of Equation 6.83 by $\sin(m\pi \xi)$:

$$(\xi - 1) \sin(m\pi \xi) = D_1 \sin(\pi \xi) \sin(m\pi \xi) + \cdots$$
$$+ D_n \sin(n\pi \xi) \sin(m\pi \xi) + \cdots. \tag{6.84}$$

We can now integrate both sides of Equation 6.84 with respect to ξ, over the interval $0 \leq \xi \leq 1$ for which the function is defined:

$$\int_0^1 (\xi - 1) \sin(m\pi \xi)\, d\xi = D_1 \int_0^1 \sin(\pi \xi) \sin(m\pi \xi)\, d\xi$$
$$+ D_2 \int_0^1 \sin(2\pi \xi) \sin(m\pi \xi)\, d\xi + \cdots. \tag{6.85}$$

From the definition of orthogonality (Equation 6.81), we know that all the integrals on the right-hand side of Equation 6.85 will be identically zero *except* the one integral for which $m = n$; as a result, we can write Equation 6.85 more simply as:

$$\int_0^1 (\xi - 1) \sin(n\pi \xi)\, d\xi = D_n \int_0^1 \sin^2(n\pi \xi)\, d\xi. \tag{6.86}$$

Equation (6.86) can be rearranged to give us a formula for our constants:

$$D_n = \frac{\int_0^1 (\xi - 1)\sin(n\pi\xi)\,d\xi}{\int_0^1 \sin^2(n\pi\xi)\,d\xi}. \tag{6.87}$$

In order to find an explicit statement of our coefficients, we have to perform the indicated integrations. The integral in the numerator evaluates as

$$\int_0^1 (\xi - 1)\sin(n\pi\xi)\,d\xi = \int_0^1 \xi\sin(n\pi\xi)\,d\xi - \int_0^1 \sin(n\pi\xi)\,d\xi \tag{6.88}$$

$$\int_0^1 (\xi - 1)\sin(n\pi\xi)\,d\xi = \frac{\sin(n\pi\xi)}{(n\pi)^2} - \frac{\xi\cos(n\pi\xi)}{n\pi}\Big|_0^1 + \frac{\cos(n\pi\xi)}{n\pi}\Big|_0^1 \tag{6.89}$$

$$\int_0^1 (\xi - 1)\sin(n\pi\xi)\,d\xi = -\frac{1}{n\pi}, \tag{6.90}$$

while the integral in the denominator is

$$\int_0^1 \sin^2(n\pi\xi)\,d\xi = \left[\frac{\xi}{2} - \frac{1}{4n\pi}\sin(2n\pi\xi)\right]_0^1 \tag{6.91}$$

$$= \frac{1}{2}. \tag{6.92}$$

Substituting Equations 6.90 and 6.92 into Equation 6.87 gives us the formula for the coefficients of the Fourier series:

$$D_n = -\frac{2}{n\pi}. \tag{6.93}$$

6.6 Evaluating the solution

The final step in obtaining a solution to the Augean Stables problem is to assemble the pieces we have found in the preceding sections. From Equations 6.79 and 6.93 we can see that

$$\theta_t = -\frac{2}{\pi}\sum_{n=1}^{\infty} \frac{e^{-(n\pi)^2\tau}\sin(n\pi\xi)}{n}. \tag{6.94}$$

(Note we have factored the constant $-2/\pi$ from the D_n and brought it outside the summation.) This is the transient portion of the solution; however, we still need to add in the steady-state part that we pulled out in Section 6.5.2. Combining these solutions yields the complete solution of the governing equation:

$$\theta = 1 - \xi - \frac{2}{\pi}\sum_{n=1}^{\infty} \frac{e^{-(n\pi)^2\tau}\sin(n\pi\xi)}{n}. \tag{6.95}$$

6.6.1 Behavior in the limits

The solution given in Equation 6.95 is significantly more complicated-looking than the other solutions we have worked with in this text. Because we can't necessarily tell at a glance what the behavior of the solution will be, we have a proportionately greater responsibility to examine the solution carefully to determine if it acts in a physically realistic manner and accords with our intuition of the system. As always, the first thing to do is to check to make certain that the equation obeys the boundary conditions; we do this by setting $\xi = 0$, which gives us $\theta(\xi = 0) = 1$, and $\xi = 1$, which gives us $\theta(\xi = 1) = 0$, which is as expected.

The next thing we want to do is look at the behavior in the limits as $\tau \to \infty$ and $\tau \to 0$. The case when $\tau \to \infty$ is an easy one: as τ becomes larger and larger in Equation 6.95, the numerators of all the terms in the summation go to zero; therefore, in the limit as $\tau \to \infty$, θ will reduce to

$$\theta = 1 - \xi, \tag{6.96}$$

which is the steady-state solution we expected on the basis of our experience with the Elysian Fields problem. So, our model result accords with our physical intuition as far as the long-term results go.

We would also like to examine the equation to see if it satisfies the initial condition. In this situation, it is a little difficult to see clearly that the initial condition is satisfied everywhere in the model domain (we can see it is satisfied at the boundaries because we already checked that the equation satisfies the boundary conditions). However, we can "spot check" some values of ξ to see if they give a value of zero at $\tau = 0$. A particularly easy one to check (at least relatively speaking) is $\xi = 1/2$. When $\xi = 1/2$ and $\tau = 0$, the series reduces to

$$\sum_{n=1}^{\infty} \frac{\sin\left(\frac{n\pi}{2}\right)}{n} = 1 + 0 - \frac{1}{3} + 0 + \frac{1}{5} + \cdots \tag{6.97}$$

$$\sum_{n=1}^{\infty} \frac{\sin\left(\frac{n\pi}{2}\right)}{n} = 1 - \frac{1}{3} + \frac{1}{5} - \frac{1}{7} + \cdots . \tag{6.98}$$

This series is equal to (Gradshteyn and Ryzhik, 2000)

$$\sum_{k=1}^{\infty} \frac{(-1)^{k+1}}{(2k-1)} = \frac{\pi}{4}. \tag{6.99}$$

If we plug the value $\pi/4$ in for the infinite sum in Equation 6.95, we can see that $\theta(\xi = 1/2, \tau = 0) = 0$, so that point, at least, satisfies the initial condition. Of course, there might be something special about the point $\xi = 1/2$, so we should check a few other values, but this does give us confidence that our equation is acting as we expect.

There are no parameters in Equation 6.95, so we can't examine the way the model behavior changes as some parameter goes to 1 or 0 or ∞, as we did with the water quality model of Chapter 3. At this point, we have probably made all the "quick-and-dirty" checks on our model's behavior that we can reasonably make. Our next task will be to plot up the solution and see what it looks like for various values of τ.

6.6.2 Evaluating the solution

Since this is the first analytical solution we have developed in this text that is not "closed form" (an analytical expression that can be evaluated in a finite number of operations), it will be worthwhile to examine the way in which Equation 6.95 provides an answer to our problem. As with all equations, we will typically want to know the value of θ (our dependent variable) for some pair of (ξ, τ) (the independent variables). We therefore substitute the ξ and τ of interest into the summation in Equation 6.95, and begin adding up the terms of the sum, starting with $n = 1$ (as indicated by the limits on the sum). We continue to add terms until the absolute value of the most recent term is less than some predetermined value. This predetermined value is called the "convergence criterion," and when we have added sufficient terms that the magnitude of the most recent term is less than the convergence criterion we say the solution "has converged." (Note we use the *absolute value* of the term; the reason is that any negative term—every other term in Equation 6.95—is less than any positive convergence criterion.) We choose the convergence criterion to be small compared to the precision we require in our calculated value of θ. For example, if we would like our estimate of θ to have a precision of ± 0.01, we might choose our convergence criterion to be ≤ 0.005.

Once the infinite series converges, we multiply the sum by $-\pi/2$, and add that value to $1 - \xi$; the result of that calculation is the value of θ for the selected (ξ, τ). Of course, we will usually be interested in the values of θ for a given value of τ on an interval $0 \leq \xi \leq 1$, so we will calculate θ repeatedly at the τ of interest for $\xi = 0.001$, $\xi = 0.002$, and so on (or similar values of ξ, depending on the resolution we need along the ξ-axis). A plot of the solution on the domain $0 \leq \xi \leq 1$ for a selection of τ is shown in Figure 6.4.

Examination of Figure 6.4 shows that our solution behaves in a physically reasonable fashion. At the initial time $(\tau = 0)$, everywhere in the domain $\theta = 0$ except at the boundary $\xi = 0$ (this is the initial condition, and is not visible in the figure because the solution coincides with the axes). As time passes, head increases across the domain, rising first (and most rapidly) near the fixed-head boundary at $\xi = 0$, and gradually progressing toward the fixed-head boundary at $\xi = 1$. Eventually, the head profile nears steady state, and by $\tau = 1$ the head profile is very near the steady-state solution $\theta = 1 - \xi$. Note that the change is most rapid for early times (times near the instantaneous perturbation at the $\xi = 0$ boundary), and gradually slows as the head profile nears equilibrium. This is similar to the behavior we have noted in our other transient problems, where the dependent variable goes toward steady-state asymptotically; again, this accords with our

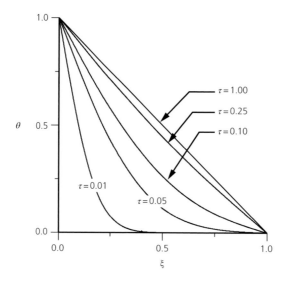

Figure 6.4 Plot of dimensionless head (θ) as a function of dimensionless position (ξ) for selected values of dimensionless time (τ) as indicated in the figure.

physical intuition of the problem, and gives us confidence that our solution is correctly representing the conceptual model.

It's interesting to ask what other quantities we can estimate using the solution given by Equation 6.95. For example, we might like to find an expression for the flux as a function of ξ and τ. From Darcy's law, we know that flux is

$$q = -K\frac{dH}{dx}. \tag{6.100}$$

Using Equations 6.25, 6.26, 6.29, and 6.31, we can nondimensionalize Equation 6.100 as:

$$q = -\frac{K(H_0 - H_L)}{L}\frac{d\theta}{d\xi}. \tag{6.101}$$

Rearranging gives

$$\frac{qL}{K(H_0 - H_L)} = -\frac{d\theta}{d\xi}. \tag{6.102}$$

From Equation 6.102, it is clear that the dimensionless flux, v, is equal to

$$v = \frac{q}{-K\frac{(H_L - H_0)}{L}} = -\frac{d\theta}{d\xi}. \tag{6.103}$$

In other words, the characteristic flux is the linearized Darcy flux—resulting from a linear gradient. This should not be a surprise, since we found a very similar result when we nondimensionalized Darcy's law in the Bandurraga Basin model of Chapter 2. If we apply Equation 6.103 to Equation 6.95, we find the equation for flux is

$$v = -\frac{d\theta}{d\xi} = -\frac{d}{d\xi}\left[1 - \xi - \frac{2}{\pi}\sum_{n=1}^{\infty}\frac{e^{-(n\pi)^2\tau}\sin(n\pi\xi)}{n}\right]. \qquad (6.104)$$

Since Fourier series are uniformly convergent (Wylie and Barrett, 1995), we can differentiate termwise to find:

$$v = 1 + 2\sum_{n=1}^{\infty}e^{-(n\pi)^2\tau}\cos(n\pi\xi). \qquad (6.105)$$

(You should do the differentiation yourself to verify that I haven't made a mistake.) Note that, as dimensionless time goes to infinity, the equation for flux goes to one. This is, of course, correct, because we expect the dimensionless flux to be unity for a linear gradient, based on both the mathematics and our experience with the Elysian Fields problem.

6.6.3 Dimensionless time

Before we move on from the analytical solution to the Augean Stables problem, I would like to examine briefly the form of the dimensionless time variable τ. Depending on whether one prefers the storativity/transmissivity notation or the diffusivity notation, the dimensionless time can be written as:

$$\tau = \frac{t}{t_c} = \frac{tD}{L^2} = \frac{tT}{SL^2}. \qquad (6.106)$$

This particular scaling of dimensionless time shows up in many hydrologic and heat transfer problems. In fact, it appears so often in heat transfer problems that scientists and engineers working in heat transfer have a special name for it: the "Fourier number," and they commonly represent it using the symbol F_o. I personally have a preference for representing dimensional quantities with the English alphabet, and dimensionless quantities with Greek letters,[9] and I will continue to use this convention throughout this text. However, it is the essence, not the nomenclature, that is important here, and you should be able to recognize the Fourier number regardless of the symbols used.

Back in Section 2.6, you may remember that we discussed calibration of the exponential decay model by estimating the time at which initial discharge fell to $1/e \approx 0.37$ (~37% of the initial discharge). This works because, no matter when you take your initial discharge measurement, discharge will decrease to $1/e \approx 0.37$ in one characteristic time unit, $1/e^2 \approx 0.14$ after two characteristic time units, $1/e^3 \approx 0.05$ after three characteristic time units, and so on. As a result, in a system characterized by pure exponential decay, the system will be 95% of the way to its "final" (equilibrium) value after three characteristic time units ($\tau = F_o = 3$), and 99.3% of the way to equilibrium when $\tau = F_o = 5$.

This type of information can be important when selecting equipment for laboratory or field tests. For example, when choosing thermocouples for making temperature measurements, the investigator will want to know how long a particular thermocouple takes to come to equilibrium with its surroundings. Manufacturers will generally provide either a thermocouple's characteristic time or the "response time" (usually the time it takes to achieve $\tau = 5$, or 99.3% of equilibrium). A thermocouple with a 15 second characteristic time (a small epoxy-beaded thermocouple or similar) will require about 01 : 15 (mm : ss) to come to approximate equilibrium with its surroundings, whereas a large, armored soil-penetration probe may have a characteristic time of 45 seconds or more, meaning it will take about 4 minutes to come to equilibrium. This is a small issue if a few dozen measurements are to be made in the course of a day, but if many hundreds or thousands of measurements are to be made, having a thermocouple that comes to equilibrium quickly will be a real asset in the field (or laboratory), potentially saving many hours of fieldwork. For additional issues related to the equilibration time of thermocouples, see Fairley and Zakrajsek (2007).

As can be seen from an examination of Figure 6.4, the Augean Stables model comes to equilibrium much more rapidly than a pure exponential decay problem. Nevertheless, the Fourier number gives us a guideline to thinking about an approach to equilibrium. If we know or can estimate a characteristic time for a problem of interest, we can often get a good feel for how long the system should take to come to a tolerable approximation of equilibrium. For example, a 50 km thick chunk of the Earth's crust (i.e., the Nazca plate (Soudoudi et al., 2011)), with a thermal diffusivity of about 2×10^{-7} m^2/s (which is about 6×10^{-6} km^2/yr), will have a characteristic time of about 400,000 years; it will therefore take approximately 1.6 million years for a piece of subducting slab to reach 95% of thermal equilibrium as it tears off and plunges into the mantle. Similarly, when pumping a well for sampling, one should pump around three to five well volumes to achieve 95–99% formation water in the outflow. (To see where this second example came from, review the water quality model of Chapter 3 and imagine the tank in Figure 3.2 is tall and thin, similar to a well-bore.)

6.7 Transient finite differences

By this time, you probably have enough experience with finite difference operators that you could write a finite difference version of the governing equation of the Augean Stables problem. However, we will add a few twists to the most basic possible difference equation; these twists will add somewhat to the complexity of the problem, but will increase numerical stability and the versatility of our method.

6.7.1 The explicit scheme

The nondimensional version of our governing equation is

$$\frac{\partial^2 \theta}{\partial \xi^2} = \frac{\partial \theta}{\partial \tau}. \tag{6.107}$$

On the basis of our previous experience with finite differences, we can write a basic finite difference version of the governing equation as

$$\frac{\theta^i_{j+1} - 2\theta^i_j + \theta^i_{j-1}}{(\Delta \xi)^2} = \frac{\theta^{i+1}_j - \theta^i_j}{\Delta \tau}, \tag{6.108}$$

where the superscript i refers to the present timestep, $i + 1$ is the future timestep, and $j - 1$, j, and $j + 1$ are the spatial indices of the current node (j) and the two flanking nodes. Furthermore, $\Delta \tau$ is the size of the timestep and $\Delta \xi$ is the distance between grid nodes; note that we have assumed the spacing between grid nodes is constant and everywhere equal. Rearranging, we can find an expression for the current node at the future timestep:

$$\theta^{i+1}_j = \left(\frac{2\Delta \tau}{\Delta^2 \xi} + 1\right)\theta^i_j - \left(\frac{\Delta \tau}{\Delta^2 \xi}\right)\left[\theta^i_{j-1} + \theta^i_{j+1}\right]. \tag{6.109}$$

Equation 6.109 is called an "explicit" finite difference scheme, because the value of a node (θ_j) at the future timestep is completely determined by the values of the nodes at the present timestep. Using Equation 6.109, we can go through all the nodes of our grid, calculating the values of θ for every node at the next timestep, thus stepping our way through the entire calculation until we reach the time we are interested in (or until the difference in all θ_j between one timestep and the next is reduced below some predetermined threshold, at which time we say the problem has reached steady state).

The explicit finite difference scheme is the easiest transient scheme to understand; because it is intuitive, it is quite appealing to beginning modelers. Unfortunately, the explicit scheme suffers from a hidden problem: in order to maintain a stable solution, the parameter $\Delta \tau / (\Delta \xi)^2$ (or, in dimensional terms, $T\Delta t / S(\Delta x)^2$) must be smaller than 0.5 for a 1D problem, and smaller than 0.25 for a 2D problem. This means that either the grid must be discretized relatively coarsely or the timestep must be relatively small. For example, if we were using a nodal spacing of $\Delta \xi = 0.1$ for the Augean Stables problem (a 1D problem), the largest timestep we could use would be $\Delta \tau = 0.005$. Because a fine grid is usually needed to represent steep gradients at times near $\tau = 0$ (remember the finite difference approximation is only valid for "small" $\Delta \xi$), explicit finite difference schemes are generally limited to very small timesteps—which makes them of limited utility for most real-life problems.

6.7.2 Fully implicit finite differences

In order to overcome the limitations of the explicit method, we need to explore other possible finite difference schemes. In the explicit scheme, we have evaluated all the θ_j at the present timestep to find the value of θ_j at the future timestep (this is sometimes called a "backward difference" scheme). Alternatively, we could evaluate the θ_j at the future timestep:

$$\frac{\theta_{j+1}^{i+1} - 2\theta_j^{i+1} + \theta_{j-1}^{i+1}}{(\Delta\xi)^2} = \frac{\theta_j^{i+1} - \theta_j^i}{\Delta\tau}. \tag{6.110}$$

This is a "forward difference" scheme, or *fully implicit* finite difference method, because none of the values of the nodes are known explicitly (except for θ_j^i). As a result, all of the values of θ_j must be calculated in some fashion—either iteratively, or using some type of matrix solver routine (i.e., Gaussian row reduction) at each timestep.

How can we solve for the unknown θ_j simultaneously? The first step is to solve Equation 6.110 to get the known and unknown quantities on opposite sides of the equation. Keeping the unknowns and their coefficients on the left-hand side, and the known quantities on the right-hand side:

$$\frac{\Delta\tau}{(\Delta\xi)^2}\theta_{j-1}^{i+1} + \left[1 - \frac{2\Delta\tau}{(\Delta\xi)^2}\right]\theta_j^{i+1} + \frac{\Delta\tau}{(\Delta\xi)^2}\theta_{j+1}^{i+1} = \theta_j^i, \tag{6.111}$$

If the model domain is discretized into $N + 2$ nodes (N computational nodes, plus two nodes, $j = 0$ and $j = N + 1$, representing head boundaries at $\xi = 0$ and $\xi = 1$), Equation 6.111 defines a set of N equations with N unknowns. This system can be written as a matrix equation:

$$\begin{bmatrix} -1 - \frac{2\Delta\tau}{(\Delta\xi)^2} & \frac{\Delta\tau}{(\Delta\xi)^2} & 0 & \cdots & 0 \\ \frac{\Delta\tau}{(\Delta\xi)^2} & -1 - \frac{2\Delta\tau}{(\Delta\xi)^2} & \frac{\Delta\tau}{(\Delta\xi)^2} & \cdots & 0 \\ 0 & \frac{\Delta\tau}{(\Delta\xi)^2} & -1 - \frac{2\Delta\tau}{(\Delta\xi)^2} & \cdots & 0 \\ \vdots & \vdots & \vdots & \vdots & \vdots \\ 0 & 0 & \cdots & \frac{\Delta\tau}{(\Delta\xi)^2} & -1 - \frac{2\Delta\tau}{(\Delta\xi)^2} \end{bmatrix} \begin{bmatrix} \theta_{j=1}^{i+1} \\ \theta_{j=2}^{i+1} \\ \theta_{j=3}^{i+1} \\ \vdots \\ \theta_{j=N}^{i+1} \end{bmatrix}$$

$$= \begin{bmatrix} -\theta_{j=1}^i - \frac{\Delta\tau}{(\Delta\xi)^2}\theta_{j=0}^i \\ -\theta_{j=2}^i \\ -\theta_{j=3}^i \\ \vdots \\ -\theta_{j=N}^i - \frac{\Delta\tau}{(\Delta\xi)^2}\theta_{j=N+1}^i \end{bmatrix}. \tag{6.112}$$

The matrices in Equation 6.112 are for the N equations and N unknowns, where the current timestep is i and the future timestep is $i + 1$. At the initial time,

$i = 0$, all the knowns $j = 1 \cdots N$ on the right-hand side are zero, because the initial conditions are zero. $\theta_{j=0}$ represents the boundary at $\xi = 0$, and so is equal to one for all i. At timesteps $i > 0$, the matrix of knowns (on the right-hand side) is filled up with the boundary conditions and known values of θ from the previous timestep. If all this seems confusing (I would be surprised if it *didn't* seem confusing), try writing out a few of the equations by hand to see how the matrix is filled up.

For most problems, we would solve our system of equations by filling up a series of arrays that represent the coefficients and the "knowns" (the right-hand side), and call a matrix solver package or library functions. In the present instance, however, we are working on a 1D problem, and the resulting coefficient matrix is tridiagonal. You may recall that we discussed this in Section 4.8.2, where we mentioned that tridiagonal matrices could be solved by the Thomas algorithm, and that applies to the present case as well. As a result, 1D problems, at least, can be solved without the use of proprietary (and therefore expensive) solver packages.

6.7.3 A generic difference formulation

We presented a fully explicit finite difference scheme in Section 6.7.1 and a fully implicit finite difference scheme in Section 6.7.2. If we were going to write a computer program to calculate our numerical approximations for us (e.g., a Fortran 90/95/2003 or C/C++ program), it would be nice if we could write the code in such a way that we could choose at run time whether we wanted to use an explicit formulation, an implicit formulation, or perhaps something in between. In fact, there is a way we can do this; consider a third version of our finite difference representation of the Augean Stables governing equation,

$$\alpha \frac{\theta_{j-1}^{i+1} - 2\theta_j^{i+1} + \theta_{j+1}^{i+1}}{(\Delta\xi)^2} + (1 - \alpha) \frac{\theta_{j-1}^i - 2\theta_j^i + \theta_{j+1}^i}{(\Delta\xi)^2} = \frac{\theta_j^{i+1} - \theta_j^i}{\Delta\tau}. \tag{6.113}$$

For the finite difference version of the governing equation presented in Equation 6.113, if we choose $\alpha = 1$, the equation reduces to the fully implicit formulation; on the other hand, if we let $\alpha = 0$, we get an explicit finite difference representation of our governing equation. We can even pick α such that we get a weighted average of the two schemes. For example, if we pick $\alpha = 0.5$ we get an equally weighted average of the two methods known as the *Crank-Nicolson method* (Wang and Anderson, 1982).

Although the formulation in Equation 6.113 probably looks significantly more complex than either the explicit or fully implicit schemes, the method of solution is essentially the same as for the fully implicit method: bring all the knowns to one side of the equation and all the unknowns to the other side, set up a matrix equation of the form

$$[A][\theta] = [R], \tag{6.114}$$

where [A] is the $N \times N$ coefficient matrix, [θ] is the $1 \times N$ matrix of unknowns, and [R] is the $1 \times N$ matrix of knowns, which includes boundary conditions, values of θ from the prior (current) timestep, and any source/sink terms. The system of equations is then solved using some preferred method. Once the unknowns are determined for the future timestep, the timestep counter is updated, the new values of θ are transferred to the array for the current timestep, and the process is repeated until the target time (or steady state) is reached.

6.8 Conclusions

In this chapter, we have examined the transient diffusion equation—perhaps the most versatile of the three potential flow equations we have investigated to date (i.e., the Laplace equation, Poisson's equation, and the diffusion equation). In fact, the diffusion equation is a generalization of the other two equations; the Laplace equation is derived from the diffusion equation by the specialization of allowing the temporal derivative to go to zero (and thus representing the steady-state solution to the diffusion equation). The Poisson equation also represents steady-state behavior (i.e., the temporal derivative of the transient diffusion equation equals zero), but with the inclusion of a steady source term, as can be seen from the general (nonhomogeneous) form of the transient diffusion equation:

$$\nabla^2\theta + R = \frac{\partial\theta}{\partial\tau},$$ (6.115)

where R is a sink/source term.

In addition to using the diffusion equation to investigate the impact of a sudden rise in water level at the boundary of a confined aquifer, we introduced the method of separation of variables to obtain an analytical solution to the governing PDE. Although it won't work for every problem, separation of variables is a powerful and versatile method for solving many PDEs, and it is well worth the time and effort it takes to master this approach. While working through the separation of variables method, we introduced the concept of orthogonality—another powerful and important mathematical concept, with applications far beyond the solution of PDEs.

Finally, we developed a finite difference version of the governing PDE for transient diffusion, and extended the basic (explicit) formulation to cover both a fully implicit finite difference scheme and a weighted average of the explicit and implicit difference methods. Although intuitive, the explicit difference scheme is of limited usefulness because small timesteps are usually required to maintain numerical stability. Fortunately, the implicit difference method avoids this limitation, although it requires somewhat greater mathematical/computational sophistication to achieve a solution using this method. Solving the resulting system of equations is relatively easy for 1D equations using the Thomas algorithm for tridiagonal matrices (see Appendix C), but methods for higher dimensional problems are also available; see, for example, Patankar (1980) or Press et al. (1992).

6.9 Problems

1. Plot the analytical solution of the Augean Stables problem (Eq. 6.95) at times $\tau = 0.01$, $\tau = 0.05$, $\tau = 0.1$, $\tau = 0.2$, $\tau = 0.5$, and $\tau = 1.0$. Use a spacing of $\Delta \xi = 0.01$ to get a nice curve for your plots.
2. Plot the numerical solution for flux at the Augean Stables (Eq. 6.105) at times $\tau = 0.01$, $\tau = 0.05$, $\tau = 0.1$, $\tau = 0.2$, $\tau = 0.5$, and $\tau = 1.0$. Use a spacing of $\Delta \xi = 0.01$ to compare with the previous problem.
3. Using orthogonality, calculate the coefficients A_n and B_n in the Fourier series,

$$f(t) = A_0 + \sum_{n=1}^{\infty} A_n \sin(2n\pi t) + B_n \cos(2n\pi t), \tag{6.116}$$

for the function

$$f(t) = t, \tag{6.117}$$

where the function $f(t)$ is periodic on the interval from 0 to 1.
4. Use separation of variables to find a solution to the governing equation,

$$\frac{\partial^2 \theta}{\partial \xi^2} + \frac{\partial^2 \theta}{\partial \zeta^2} = 0, \tag{6.118}$$

with boundary conditions

$$\theta(\xi = 0, \zeta) = 0, \tag{6.119}$$

$$\theta(\xi = 1, \zeta) = 0, \tag{6.120}$$

$$\theta(\xi, \zeta = 0) = 0, \tag{6.121}$$

$$\theta(\xi, \zeta = 1) = 2\xi(1 - \xi). \tag{6.122}$$

Before you actually attempt to do the math, it would be instructive to stop and ask yourself what you think the solution *should* look like.
5. For an advanced challenge: Although it may be difficult to include time-varying terms in an analytical solution, including time-varying boundary conditions or source terms in a numerical solution is mostly an exercise in bookkeeping. If your programming skills are up to it, try solving the Augean Stables problem,

$$\frac{\partial^2 \theta}{\partial \xi^2} = \frac{\partial \theta}{\partial \tau}, \tag{6.123}$$

with a time-varying boundary condition

$$\theta(\xi, \tau = 0) = 0, \tag{6.124}$$

$$\theta(\xi = 0, \tau) = \epsilon \sin(\omega \tau), \tag{6.125}$$

$$\theta(\xi = 1, \tau) = 0, \tag{6.126}$$

where ϵ is the height of the change in water level at the River Alpheus, and ω is a parameter (the *angular frequency*) that controls how rapidly the boundary condition cycles. If your programming skills aren't up to the challenge, you can at least try writing out the matrices for the fully implicit solution to see how time-varying conditions can be incorporated into the finite difference formulation.

Notes

1 Although this was a terrible thing to do, it pales in comparison to some of the other things the Greeks did. For example, Atreus, father of Agamemnon, found out that his wife Aerope was cheating on him with his twin brother Thyestes. In retaliation, he killed Thyestes's children and fed them to him for dinner. Thyestes's son (by his own daughter) Aegisthus later murdered Atreus in revenge. One would hope that Agamemnon might have learned something from this bad example, but, sadly, that was not the case. Among other atrocities, Agamemnon sacrificed his own daughter Iphigenia to the goddess Artemis, in an attempt to gain a fair wind for sailing.

2 For a fascinating and very readable account of a scientific investigation of the Oracle of Delphi, see Broad (2006).

3 Dante places Cerberus in the third circle of Hell (the Gluttons), where he rips and tears at the sinners, sunk in a disgusting, frozen muck of black snow and dirty water; see Ciardi (1982), Canto VI.

4 This is known as the *Boussinesq approximation*: that density differences are small enough to be neglected, except for the storage term (or, when considering buoyancy, in terms where the density differences are multiplied by gravitational acceleration).

5 For transient problems, the initial condition is the preferred nonhomogeneous boundary condition. In steady problems, the nonhomogeneous condition is the one about which we are going to construct a Fourier series with the remaining integration constant.

6 We choose the constant to be squared because squared numbers are always positive—at least on the real number line—and thus there is no ambiguity regarding the sign of the (squared) number.

7 It is probably technically more correct to say the trivial solution is *always* part of the solution, since zero can be added to any function without changing it—so the trivial solution is always added to the final solution. But this is splitting hairs.

8 This is often the case when a mathematical physicist invents new mathematical tools to solve physical problems. In general, it takes pure mathematicians years—sometimes decades—to work out rigorous proofs for the tools the applied mathematicians develop. The electrical engineer and physicist Oliver Heaviside (1850 – 1925) suffered similar problems in winning acceptance of his operational calculus.

9 I was first introduced to this convention by Professor P.J. Pagni, my heat transfer professor at UC Berkeley.

References

Bell, E.T. (1937) *Men of Mathematics*, Simon and Schuster, New York.

Broad, W.J. (2006) *The Oracle: The Lost Secrets and Hidden Message of Ancient Delphi*, The Penguin Press, New York.

Ciardi, J. (1982) *The Inferno, by Dante Aleghieri; A Verse Rendering for the Modern Reader*, New American Library, New York.

Fairley, J.P. and Zakrajsek, J.R. (2007) A physical antialias filter for time-series temperature measurements. *Ground Water Monitoring & Remediation*, **27** (1), 103–107.

Farlow, S.J. (1993) *Partial Differential Equations for Scientists and Engineers*, Dover Publications, New York.

Freeze, R.A. and Cherry, J.A. (1979) *Groundwater*, Prentice Hall, Englewood Cliffs.

Gradshteyn, I.S. and Ryzhik, I.M. (2000) *Table of Integrals, Series, and Products*, 6th edn., Academic Press, New York.

Hermance, J.F. (1999) *A Mathematical Primer on Groundwater Flow*, Prentice Hall, Englewood Cliffs.

Lathi, B.P. (1998) *Signal Processing and Linear Systems*, Oxford University Press, Oxford.

Logan, J.D. (2004) *Applied Partial Differential Equations*, 2nd edn., Springer Science+Business Media, LLC, New York.

Logan, J.D. (2006) *A First Course in Differential Equations*, Springer Science+Business Media, LLC, New York.

Patankar, S.V. (1980) *Numerical Heat Transfer and Fluid Flow*, Routledge, Taylor & Francis Group, New York.

Press, W.H., Flannery, B.P., Teukolsky, S.A., and Vetterling, W.T. (1992) *Numerical Recipies in Fortran 77: The Art of Scientific Computing*, Cambridge University Press, Cambridge.

Soudoudi, F., Yuan, X., Asch, G., and Kind, R. (2011) High resolution image of the geometry and thickness of the subducting Nazca lithosphere beneath northern Chile. *Journal of Geophysical Research*, **116**, B04302, doi:10.1029/2012JB007829.

Stillwell, J. (2002) *Mathematics and its History*, 2nd edn., Springer Science+Business Media, LLC, New York.

Wang, H.F. and Anderson, M.P. (1982) *Introduction to Groundwater Modeling: Finite Difference and Finite Element Methods*, Academic Press, New York.

Wylie, C.R. and Barrett, L.C. (1995) *Advanced Engineering Mathematics*, 6th edn., McGraw-Hill, Inc., New York.

CHAPTER 7

The Theis equation

Chapter summary

In this chapter, we examine the problem of pumping a well located in an areally extensive aquifer. This problem was first examined by C.V. Theis in his classic paper—a paper that ushered in a new era of quantitative analysis of groundwater hydraulics (Theis, 1935). Despite the fundamental nature of this model and its impact on hydrogeology, few hydrologists take the time to look carefully at the mathematics behind the model: how the equation is derived, what assumptions are behind it, and how it is solved. This lack of attention is unfortunate, not only because of the importance of the model to groundwater hydrology, but also because the model has much to teach hydrologists about the nature of groundwater flow, the relationship between time and space in the vicinity of a pumping well, and the robustness of mathematical models applied to physical systems. In our examination of this problem, we will place our pumping well within the confines of yet another semi-mythical field of dreams: the Plain of La Mancha, nominally the largest plateau of the Iberian Peninsula, but for our purposes located within the infinitely vast regions of the ideal.

7.1 The Knight of the Sorrowful Figure

How many of you are familiar with the story *El Ingenioso Hidalgo de don Quijote de la Mancha* or, in English, *The Ingenious Gentleman Don Quixote of La Mancha*? This famous work by Miguel de Cervantes, written in 1605 (Part 1) and 1615 (Part 2), was perhaps the first modern novel, forming a bridge from the chivalric tradition. The literature of chivalry was intended to teach European nobles some manners and divert them from the popular pastime of abusing those over whom they had authority. Chivalric stories are largely disjointed anecdotes intended to provide a moral standard. In contrast, the modern novel primarily focuses on character development and the hero's psychological growth. Cervantes's novel, usually referred to simply as *Don Quixote* for short, cleverly bridges the distance between these two genres by building the story around the main character, don Quijano (who later christens himself don Quixote, or "the great sir Quijano"),

Models and Modeling: An Introduction for Earth and Environmental Scientists, First Edition. Jerry P. Fairley.
© 2017 John Wiley & Sons, Ltd. Published 2017 by John Wiley & Sons, Ltd.
Companion website: www.wiley.com/go/Fairley/Models

whose reason has become unseated from reading too many books of chivalry and knighthood. (The appellative "don" is the Spanish equivalent of "sir," indicating knighthood. More recently, this title has been applied broadly to mean "gentleman," similar to the term "caballero," an honorific in Latin American Spanish that literally translates as "horseman.") Don Quixote spends most of the novel (actually two novels, written about 10 years apart, but now generally published together) sallying forth on adventures with his would-be squire, Sancho Panza, and his horse, Rocinante. Usually, these adventures involve his meddling (often violently) in affairs that do not concern him, with unpleasant consequences for himself and, more commonly, for Sancho.

Probably, the most famous of don Quixote's adventures is his attack on the windmills of the Plain of La Mancha, at the beginning of his second quest (Don Quixote part 1, chapter VIII). Don Quixote attacks the windmills under the misapprehension that they are giants, with disastrous results. Sancho Panza makes clear the use to which this wind power is put: "'Take care, sir,' cried Sancho. 'Those over there are not giants, but windmills. Those things that seem to be their arms are sails which, when they are whirled about by the wind, turn the millstone."' Of course, we know that windmills may be put to many uses other than grinding grain. The extensive and fertile area of La Mancha is the largest plain on the Iberian Peninsula, well known for viniculture, cereals (corn, barely, wheat, sunflowers, alfalfa), and saffron, as well as goats, sheep (the home of Manchego cheese), and other stock. Some of the water for this semiarid region comes from the Guadiana, Jabalón, Zácara, Cigüala, and Júcar rivers. The plain, however, is underlain by an extensive alluvial/carbonate aquifer that provides plentiful groundwater to support agriculture. The La Mancha Oriental aquifer system, in particular, extends over 7260 km^2 on the eastern half of the La Mancha plain. It is recharged by the Júcar river, and maintains approximately 80,000 h of irrigated agricultural land and potable water for 275,000 inhabitants, with an annual pumpage of about 450 Mm^3.

7.1.1 The Groundwater of La Mancha

In the summer of 2001, I was living on the east coast of Spain, and I decided to take a trip up through the Pyrenees and down into central Spain. During my journey, I was mightily impressed with the vastness of the La Manchan Plain and the richness of the agriculture there. Apart from the beauty of the sunflowers, however, I wondered about the use of irrigation, and how the extensive exploitation of groundwater, which has been going on since Cervantes's time (at least since the 1600s), has impacted the potentiometric surface in the aquifer.

As you have no doubt begun to suspect, the narratives accompanying the problems in this text are primarily intended to set up more or less idealized realms, within which we can examine the response of groundwater systems to various situations without worrying about pesky questions of reality—for example, heterogeneity, or boundary conditions that vary in time or space. This is certainly the case with our story of don Quixote or, as he is sometimes referred to, "the Knight of the Sorrowful Figure." For us, the Plain of La Mancha will exist as a limitless, featureless plateau, underlain by an equally vast aquifer, the properties of which can only be described as "ideal." In the center of this great plain exists a windmill that has been pumping at a steady rate since the 1600s. The questions that I wish to investigate about this windmill, or rather about the aquifer beneath the windmill, are the following: is the aquifer at steady-state with respect to the pumping of this windmill? What is the shape of the cone of depression (i.e., what is the level of the potentiometric surface) that has resulted from this pumping?

7.2 Statement of the problem

We begin our investigation of the aquifer of La Mancha by restating the questions we wish to answer.

FIND: The depression (drawdown) of the potentiometric surface of the aquifer underlying the Plain of La Mancha, in response to constant-rate pumping by a well. In addition, we wish to know whether the aquifer has reached steady state after more than 400 years of pumping.

As has often been the case with our researches, we actually know relatively little about the aquifer underlying the Plain of La Mancha. Based on our previous ruminations, however, we can claim to know at least a few basic facts.

KNOWN: The Plain of La Mancha, and the underlying aquifer, are areally extensive and (according to our understanding of the problem) are impacted solely by the constant rate pumping of a single well.

Clearly, this statement of the known relies on our somewhat fanciful description of the problem. It will, however, suffice for our present use. In this case, more important than the statement of the known "facts" of the problem is the statement of assumptions.

ASSUME: A homogeneous, isotropic aquifer of infinite areal extent and constant properties. Changes in the thickness of this aquifer due to pumping (or any other reason) are assumed to be insignificant.

Finally, we must state our approach. By now, you should be familiar with the approach we will be taking.

APPROACH: Use mass balance to develop a differential equation representing the aquifer of the Plain of La Mancha. Solve the governing equation analytically (if possible) or numerically (if necessary) and use the resulting expressions to answer the questions of interest.

7.3 The governing equation

A little bit of thought will reveal that the most obvious way to go about formulating the governing equations would be to use a radial coordinate system. You probably remember that our goal is always to develop the simplest description of the system under consideration that makes sense, and this includes using symmetry to reduce the dimensionality of the problem whenever possible. In this instance, we would need a 2D coordinate system to represent the problem in Cartesian coordinates, but we can make do with a 1D representation if we employ radial coordinates. The basic conceptual layout of the problem is shown in Figure 7.1.

The actual model domain is infinite in the r direction, so the boundary of the cylinder with outer radius labeled r in Figure 7.1 is located at an arbitrary distance from the pumping well (which is centered at $r=0$), and should not be taken as representing the edge of the domain. Because the model domain is 1D radial, with the independent variable r, the CV for this problem is a little different than that of our previous models. In this case, the CV consists of a hollow cylinder, b tall, with an inner radius of r and an outer radius of $r + \Delta r$ (Figure 7.2).

The mass balance for the model is

$$M_{in} = M_{out} + \Delta M. \tag{7.1}$$

Expanding the M_{in} term, we have

$$M_{in} = q_r 2\pi r b \rho \Delta t, \tag{7.2}$$

and the M_{out} term is

$$M_{out} = q_{r+\Delta r} 2\pi (r + \Delta r) b \rho \Delta t. \tag{7.3}$$

Since this is a transient problem, we also need a statement for ΔM,

$$\Delta M = \left[\pi (r + \Delta r)^2 b - \pi r^2 b\right] \Delta(\rho\phi), \tag{7.4}$$

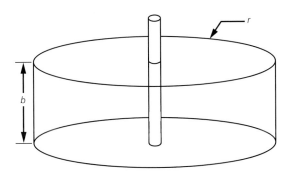

Figure 7.1 Conceptual model for a 1D radial coordinate system. The pumping well is located in the center of the coordinate system (centered on $r=0$). The thickness of the aquifer is given as b.

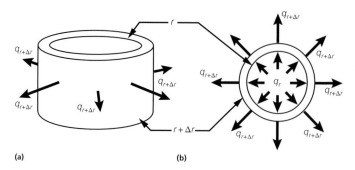

Figure 7.2 Control volume (CV) for 1D radial coordinates. Note the fluxes are drawn positive in the +r direction. (a) Perspective view of the CV. (b) Looking down from above on the CV (i.e., map view).

where ρ is the density of water and ϕ is the porosity of the aquifer. Substituting Equations 7.2, 7.3, and 7.4 into Equation 7.1 yields

$$q_r 2\pi r b \rho \Delta t = q_{r+\Delta r} 2\pi (r + \Delta r) b \rho \Delta t + \Delta(\rho\phi)\pi b \left[(r + \Delta r)^2 - r^2\right]. \tag{7.5}$$

The πb cancel in Equation 7.5, and the term in square brackets on the right-hand side expands to

$$(r + \Delta r)^2 - r^2 = 2r\Delta r + \Delta r^2. \tag{7.6}$$

Rearranging terms in Equation 7.5 therefore gives

$$2\left[(r + \Delta r)q_{r+\Delta r} - rq_r\right]\rho\Delta t = -\Delta(\rho\phi)(2r\Delta r + \Delta r^2). \tag{7.7}$$

Dividing both sides of Equation 7.7 by $\rho\Delta r\Delta t$, we have

$$2\frac{(r + \Delta r)q_{r+\Delta r} - rq_r}{\Delta r} = -\frac{(2r + \Delta r)}{\rho}\frac{\Delta(\rho\phi)}{\Delta t}, \tag{7.8}$$

which, when we take the limit as $\Delta r, \Delta t \to 0$ and cancel the factors of 2, becomes

$$\frac{\partial}{\partial r}[rq] = -\frac{r}{\rho}\frac{\partial[\rho\phi]}{\partial t}. \tag{7.9}$$

Remembering that Darcy's law is

$$q = -K\frac{dH}{dr}, \tag{7.10}$$

we can rewrite Equation 7.9 as

$$\frac{1}{r}\frac{\partial}{\partial r}\left[r\frac{\partial H}{\partial r}\right] = \frac{1}{K\rho}\frac{\partial[\rho\phi]}{\partial t}, \tag{7.11}$$

where we have used the assumption of constant properties to remove the hydraulic conductivity, K, from under the derivative. Finally, we can use the chain rule to expand the temporal derivative,

$$\frac{1}{\rho}\frac{\partial\left[\rho\phi\right]}{\partial r} = \frac{1}{\rho}\frac{d\left[\rho\phi\right]}{dH}\frac{\partial H}{\partial t}, \tag{7.12}$$

and recall the definition of storativity, S

$$S = \frac{b}{\rho}\frac{d\left[\rho\phi\right]}{dH}. \tag{7.13}$$

Substituting, we have the final dimensional form of the governing equation,

$$\frac{1}{r}\frac{\partial}{\partial r}\left[r\frac{\partial H}{\partial r}\right] = \frac{S}{T}\frac{\partial H}{\partial t}, \tag{7.14}$$

where $T = Kb$ is the transmissivity [L^2/T].

7.4 Boundary conditions

Although we now have a governing equation, we have yet to define our boundary conditions. In this instance, the boundary conditions are different from those we have encountered previously—in part because we are working in an unbounded domain, rather than one that exists within a finite range (e.g., $0 \le x \le L$ or similar). We also require an initial condition, and this is easily specified by assuming the entire domain exists at a state of equilibrium at some head H_0 for all times $t < 0$,

$$H(r, t < 0) = H_0. \tag{7.15}$$

One of our boundary conditions can be specified by prescribing that the head at a distance sufficiently far from the pumping well (i.e., located at $r \to \infty$) will always be equal to the initial head,

$$H(r \to \infty, t) = H_0. \tag{7.16}$$

The final boundary condition will be specified near $r \to 0$, where the pumping well is located. The rate at which water is pumped from the well (at an assumed constant rate of Q) will provide a specified flux boundary condition (a boundary condition of the second kind). The amount of flow through a cylinder wall of radius r can be described by Darcy's law as

$$Q = -2\pi r K b\frac{dH}{dr}. \tag{7.17}$$

This can be rearranged to

$$r\frac{dH}{dr} = -\frac{Q}{2\pi T}, \tag{7.18}$$

where T is the transmissivity $T = Kb$. This is a boundary condition of the second kind in a radial coordinate system. We could apply the condition by specifying it at some radius $r = r_w$ equal to the radius of the well. However, for our purposes, it will be more convenient to assume the radius of the well is vanishingly small compared to the domain of interest,

$$\lim_{r \to 0} \left[r \frac{dH}{dr} \right] = -\frac{Q}{2\pi T}. \tag{7.19}$$

Together, Equations 7.15, 7.16, and 7.19 comprise the initial and boundary conditions for the governing equation.

7.5 Nondimensionalization

To approach the nondimensionalization of the governing equation and initial/boundary conditions, we turn first to the conditions themselves to find characteristic quantities on which to normalize. It is clear from inspection that the head at the initial time, H_0, provides one characteristic value for normalizing the dependent variable, but there is no clear value of head with which it may be paired to derive a range. Accordingly, we assign the range to an arbitrary characteristic quantity H_c, and define the dimensionless dependent variable θ as

$$\theta = \frac{H - H_0}{H_c}. \tag{7.20}$$

Solving Equation 7.20 for H and taking the first derivative yields

$$H = H_c\theta + H_0, \tag{7.21}$$
$$dH = H_c d\theta. \tag{7.22}$$

Substituting Equation 7.22 into Equation 7.15 gives the statement

$$\frac{1}{r} \frac{\partial}{\partial r} \left[r \frac{\partial \theta}{\partial r} \right] = \frac{1}{D} \frac{\partial \theta}{\partial t}, \tag{7.23}$$

where we have used the hydraulic diffusivity,

$$D = \frac{T}{S}, \tag{7.24}$$

to give a more compact notation. To this point, the nondimensionalization procedure has not provided us with sufficient information to define the characteristic head in terms of known parameters; therefore, we move on to substituting the dimensionless head into the initial and boundary conditions. The initial condition becomes

$$\theta(r, t < 0) = 0. \tag{7.25}$$

The boundary conditions are

$$\theta(r \rightarrow \infty, t) = 0 \tag{7.26}$$

and

$$\lim_{r \to 0}\left[r\frac{d\theta}{dr}\right] = -\frac{Q}{2\pi\,TH_c}. \tag{7.27}$$

In Equation 7.27, we see the units of the quantities $Q/2\pi T$ are those of [L]; we can therefore set H_c equal to $Q/2\pi T$ to obtain

$$\lim_{r \to 0}\left[r\frac{d\theta}{dr}\right] = -1. \tag{7.28}$$

7.5.1 Normalizing independent variables the usual way

On the basis of our previous experience with nondimensionalization, we would expect to continue nondimensionalizing our independent variables by finding characteristic quantities in the auxiliary conditions against which to normalize. An examination of the boundary conditions, however, shows that there are no obvious characteristic quantities, since both r and t range from 0 to ∞. Following our nondimensional heuristic, we therefore normalize on arbitrary characteristic quantities r_c and t_c (distance and time, respectively),

$$\xi = \frac{r}{r_c}, \tag{7.29}$$

$$r = r_c\xi, \tag{7.30}$$

$$dr = r_c d\xi, \tag{7.31}$$

and

$$\tau = \frac{t}{t_c}, \tag{7.32}$$

$$t = t_c\tau, \tag{7.33}$$

$$dt = t_c d\tau. \tag{7.34}$$

Substituting into Equation 7.23 yields

$$\frac{1}{r_c^2}\frac{1}{\xi}\frac{\partial}{\partial\xi}\left[\xi\frac{\partial\theta}{\partial\xi}\right] = \frac{1}{Dt_c}\frac{\partial\theta}{\partial\tau}, \tag{7.35}$$

$$\frac{1}{\xi}\frac{\partial}{\partial\xi}\left[\xi\frac{\partial\theta}{\partial\xi}\right] = \frac{r_c^2}{Dt_c}\frac{\partial\theta}{\partial\tau}. \tag{7.36}$$

At this point, we must stop, because we are faced with a dilemma. Although the grouping of "constants" multiplying the temporal derivative on the right-hand side of the equation are correctly dimensionless (as they must be, since all the other terms in the equation are dimensionless), no characteristic quantities have appeared that we can use to set the values of r_c and t_c. Although it is tempting to think we could somehow set $r_c = \sqrt{Dt_c}$ and $t_c = D/r_c^2$, this is really a "hall of mirrors," with the values of r_c and t_c echoing back and forth forever, but with neither characteristic quantity ever being defined. In fact, our failure to nondimensionalize the independent variables hints at a hidden, but fundamental, scaling between space and time that we will explore in the following sections.

7.5.2 Finding similarity

To find an appropriate scaling for the independent variables of the governing equation, we must think more broadly about the meaning of the term r_c^2/Dt_c in Equation 7.36. In effect, this juxtaposition of "characteristic quantities" is telling us that *there are no characteristic temporal or spatial quantities for this problem, because time and space scale together.* It appears that time is inversely proportional to the square of distance and directly proportional to the diffusivity, and the spatial variable is inversely proportional to the square root of the diffusivity multiplied by time. If this is truly the case (and, in this case, it is), we can find our way out of this impasse by defining a new dimensionless variable

$$\eta = \frac{r^2}{4Dt}. \tag{7.37}$$

The new variable η is a *similarity variable*, and is so-called because it incorporates the two independent variables (in this case, time and distance) that vary in proportion one to the other. We have added a factor of 4 to the denominator of Equation 7.37 because it simplifies the following expressions. However, the mathematics works out the same whether the factor of 4 is included or not, so the reader should not feel that I have "pulled a fast one" based on prior knowledge.

To introduce the similarity variable into the governing equation (Eq. 7.23), we can use the chain rule to expand the definitions of the derivatives d/dr and d/dt,

$$\frac{d}{dr} = \frac{d}{d\eta}\frac{d\eta}{dr}, \tag{7.38}$$

$$\frac{d}{dt} = \frac{d}{d\eta}\frac{d\eta}{dt}. \tag{7.39}$$

Using the definition of η from Equation 7.37, we can find

$$\frac{d\eta}{dr} = \frac{r}{2Dt}, \tag{7.40}$$

$$\frac{d\eta}{dt} = -\frac{r^2}{4Dt^2}. \tag{7.41}$$

So, Equations 7.38 and 7.39 can be rewritten:

$$\frac{d}{dr} = \frac{r}{2Dt}\frac{d}{d\eta}, \tag{7.42}$$

$$\frac{d}{dt} = -\frac{r^2}{4Dt^2}\frac{d}{d\eta}. \tag{7.43}$$

The reader should verify that substituting Equations 7.42 and 7.43 into Equation 7.23 yields the new governing equation in terms of η:

$$\frac{d}{d\eta}\left[\eta\frac{d\theta}{d\eta}\right] + \eta\frac{d\theta}{d\eta} = 0. \tag{7.44}$$

7.5.3 The auxiliary conditions

The astute reader will note that the signs of partial differentiation that were used in Equation 7.23 have been replaced with ordinary differentiation in Equation 7.44, in keeping with our combining of two independent variables into one. The use of the similarity transform offers an enormous simplification in our representation of the problem, because an ordinary differential equation (Eq. 7.44) is, in general, much easier to solve than a partial differential equation (7.23). However, we are now faced with a new difficulty: our original governing equation required three auxiliary conditions (one initial condition and two boundary conditions) to be "properly posed." In contrast, the ordinary differential equation represented by Equation 7.44 only requires two conditions. In order to avoid having an "overdetermined" system, we must find a way of combining or dropping one of the conditions applied to the original governing equation.

To begin, we examine the nonhomogeneous boundary condition. Applying Equation 7.42 to Equation 7.28, and noting from our definition of η that $\eta \to 0$ as $r \to 0$, we arrive at a statement of this boundary condition in terms of η,

$$\lim_{\eta \to 0}\left[r\frac{r}{2Dt}\frac{d\theta}{d\eta}\right] = -1, \tag{7.45}$$

$$\lim_{\eta \to 0}\left[\frac{2r^2}{4Dt}\frac{d\theta}{d\eta}\right] = \lim_{\eta \to 0}\left[2\eta\frac{d\theta}{d\eta}\right] = -1, \tag{7.46}$$

$$\lim_{\eta \to 0}\left[\eta\frac{d\theta}{d\eta}\right] = -\frac{1}{2}. \tag{7.47}$$

Continuing on to the homogeneous conditions, we can see that, as $t \to 0$, $\eta \to \infty$, so for the initial condition we have

$$\theta(\eta \to \infty) = 0. \tag{7.48}$$

For $r \to \infty$, $\eta \to \infty$ also; therefore, the boundary condition at infinity can be written

$$\theta(\eta \to \infty) = 0. \tag{7.49}$$

The two homogeneous conditions therefore collapse into a single auxiliary condition, providing the appropriate number of conditions for the ordinary differential equation. The final form of the nondimensional governing equation and boundary conditions is

$$\frac{d}{d\eta}\left[\eta\frac{d\theta}{d\eta}\right] + \eta\frac{d\theta}{d\eta} = 0, \tag{7.50}$$

$$\theta(\eta \to \infty) = 0, \tag{7.51}$$

$$\lim_{\eta \to 0}\left[\eta\frac{d\theta}{d\eta}\right] = -\frac{1}{2}. \tag{7.52}$$

7.6 Solving the governing equation

Despite its somewhat forbidding appearance, Equation 7.50 turns out to be relatively easy to solve—although the solution is not closed-form. Using the product rule, we can expand the second-order term

$$\frac{d\theta}{d\eta} + \eta\frac{d^2\theta}{d\eta^2} + \eta\frac{d\theta}{d\eta} = 0. \tag{7.53}$$

Collecting terms and rearranging we find

$$\frac{d^2\theta}{d\eta^2} = -\left(\frac{1+\eta}{\eta}\right)\frac{d\theta}{d\eta}. \tag{7.54}$$

In this form, our governing equation is more amenable to analytical solution. Similar to our approximate solution to the Alcatraz problem (see Section 5.6.1), we can set $\chi = d\theta/d\eta$ and substitute into Equation 7.54:

$$\frac{d\chi}{d\eta} = -\left(\frac{1+\eta}{\eta}\right)\chi. \tag{7.55}$$

By direct integration, we then obtain

$$\int \frac{1}{\chi}\frac{d\chi}{d\eta}d\eta = -\int\left(\frac{1+\eta}{\eta}\right)d\eta, \tag{7.56}$$

$$\int \frac{d\chi}{\chi} = -\int\left(\frac{1}{\eta}+1\right)d\eta = -\int\frac{d\eta}{\eta} - \int d\eta, \tag{7.57}$$

$$\ln\chi = -\ln\eta - \eta + A, \tag{7.58}$$

where the constants of integration have been combined from both sides of the equation into the constant A. Raising both sides of Equation 7.58 to a power of e and replacing χ with $d\theta/d\eta$ gives

$$\frac{d\theta}{d\eta} = A\frac{e^{-\eta}}{\eta}. \tag{7.59}$$

Integrating both sides of Equation 7.59 with respect to η yields the general solution of the governing equation:

$$\int \frac{d\theta}{d\eta} d\eta = A \int \frac{e^{-\eta}}{\eta} d\eta, \tag{7.60}$$

$$\theta = A \int \frac{e^{-\eta}}{\eta} d\eta. \tag{7.61}$$

In terms of the general solution, this is as far as we can take things. The integral on the right-hand side of Equation 7.61 has no closed-form expression. However, this integral is well known to mathematicians, who call it the *exponential integral*. This is fortunate, because the exponential integral is what is known as a "tabulated function." That is, although the integral itself cannot be expressed in closed-form, the values the integral takes for particular limits of integration have been calculated very precisely and are tabulated in any standard book of mathematical tables (e.g., Abramowitz and Stegun (1972)). Thus, the exponential integral function can be used with the same facility as any other tabulated function, such as the sine, cosine, or tangent of an angle.

Although this is as far as we can take things with respect to the general solution, we can still find the specific solution associated with the auxiliary conditions (Eqs. 7.51 and 7.52). Since the integration will always go from some value of η to infinity, the upper limit of integration will always be infinity. When $\eta \to \infty$, the integral on the right-hand side of Equation 7.61 will be zero, meeting the condition prescribed in Equation 7.51. The second condition (Eq. 7.52) requires that

$$\lim_{\eta \to 0} \left[\eta \frac{d\theta}{d\eta} \right] = -\frac{1}{2}. \tag{7.62}$$

Substituting Equation 7.59 for the derivative in Equation 7.62 gives

$$\lim_{\eta \to 0} \left[\eta \frac{Ae^{-\eta}}{\eta} \right] = \lim_{\eta \to 0} \left[Ae^{-\eta} \right] = -\frac{1}{2}. \tag{7.63}$$

Since $\lim_{\eta \to 0}(e^{-\eta}) = 1$, Equation 7.63 reduces to

$$A = -\frac{1}{2}, \tag{7.64}$$

and the complete solution of Equation 7.50 is

$$\theta(\eta) = -\frac{1}{2} \int_{\eta}^{\infty} \frac{e^{-\eta'}}{\eta'} d\eta', \tag{7.65}$$

where we have used the prime notation to indicate the "dummy" integration variables are separate from the lower limit of integration.

7.7 Theis and the "well function"

Many readers will probably recognize the form of Equation 7.65 as the solution to the Theis equation. In order to find the same expression as is commonly given, however, we need to partially re-dimensionalize our solution. For historic reasons, we will substitute the variable u for η; re-dimensionalizing the θ term, we have

$$\frac{2\pi T(H - H_0)}{Q} = -\frac{1}{2} \int_u^\infty \frac{e^{-u'}}{u'} du'. \tag{7.66}$$

Rearranging gives

$$H - H_0 = -\frac{Q}{4\pi T} \int_u^\infty \frac{e^{-u'}}{u'} du'. \tag{7.67}$$

Finally, the Theis equation is usually given in terms of drawdown from a static water level. Since the original static water level is H_0, the drawdown, $\$$ [L], is given by $H_0 - H$, making the final expression,[1]

$$\$ = \frac{Q}{4\pi T} \int_u^\infty \frac{e^{-u'}}{u'} du'. \tag{7.68}$$

Unfortunately, many end-users of the Theis equation tend to feel intimidated by the integral in Equation 7.68. Instead of thinking of this as a tabulated function— no more fearsome or difficult than a sine or cosine—the uninitiated may throw up their hands and look elsewhere. As a result, it is common to hide the integral in an innocent-looking guise known to groundwater hydrologists as the "well function," $W(u)$, which is, of course, simply a pseudonym for the exponential integral. This "sanitized" Theis equation can then be written:

$$\$ = \frac{Q}{4\pi T} W(u), \tag{7.69}$$

where u is

$$u = \frac{Sr^2}{4Tt}. \tag{7.70}$$

7.7.1 Evaluating the exponential integral

Although the exponential integral cannot be evaluated in closed-form, there must of course be *some* way of calculating values of the integral for various u, else it would not be "tabulated." The workhorse of mathematical physics for evaluating these types of functions is the Taylor series. However, because there is an essential singularity at $u = 0$, the exponential integral cannot be evaluated with a simple Taylor series expansion; instead, the Taylor series for the integrand e^{-u}/u is integrated termwise and the logarithmic singularity at the origin removed.

This leads to the well-known series expansion for the exponential integral that will be familiar to most readers,

$$W(u) = -0.577216 - \ln u + u - \frac{u^2}{2 \cdot 2!} + \frac{u^3}{3 \cdot 3!} - \cdots. \tag{7.71}$$

The series defined by Equation 7.71 works well for "small" values of u (e.g., $u \ll 1$), but it requires a great many terms to obtain the necessary precision as u increases. For many practical situations, groundwater hydrologists will be working with problems for which the convergent series (Eq. 7.71) is sufficient. The groundwater modeler, on the other hand, may be interested in using the analytical solution to verify that her/his numerical model is producing correct results far from a simulated pumping well, or at very early times, in which case Equation 7.71 may not provide the required precision in a reasonable number of terms. For these cases, another series solution for the exponential integral can be developed by repeated application of integration by parts. The resulting series belongs to that class of series known as *asymptotic series*; these series are (often) divergent, but are still useful because they give excellent approximations of the function in question as the independent variable approaches some specified value (typically 0 or ∞). We use the symbol "~" (instead of "=") to indicate a series is an asymptotic representation of a particular function. Typically, we follow the series with an indication of the values for which the series gives the best approximations. For the exponential integral, the asymptotic series for large u is given by

$$W(u) \sim \frac{e^{-u}}{u}\left[1 - \frac{1}{u} + \frac{2!}{u^2} - \frac{3!}{u^3} + \cdots + \frac{n!}{(-u)^n} - \cdots\right], \quad u \to \infty. \tag{7.72}$$

In Equation 7.72, the statement $u \to \infty$ can be read as "when u is large." In this context, "large" means "significantly more than one," but the approximation improves as u gets larger.

7.8 Back to the beginning

We began our examination of the effects of pumping on the Plain of La Mancha with the following two questions: what does the potentiometric surface look like after 400 years of constant rate pumping? Has the aquifer come to equilibrium yet? We now have the tools to answer these questions.

7.8.1 Equilibrium: Are we there yet?

A common definition of equilibrium is a situation in which there is no change. More refined definitions of equilibrium are available and, in fact, there are numerous *kinds* of equilibria, for example, static and dynamic equilibria, stable and unstable equilibria, and so on. For this situation, we will take a somewhat simplistic view of equilibrium: if the potentiometric surface drops by a measurable amount over the course of an annual cycle, and if another drop can reasonably be expected to occur over the following annual cycle, then we can assume that equilibrium has not been reached.

To examine this question, we will first define a quantity $\Delta\theta$, which is the annual decline in the potentiometric surface in the neighborhood of a pumping well. Although the term "neighborhood" is somewhat nebulous, we can make it more definite by defining some radius from the well—say, 1 m. The quantity $\Delta\theta$ can be expanded in terms of Equation 7.65 as

$$\Delta\theta = \theta_a - \theta_b = -\frac{1}{2}\int_{\eta_a}^{\infty}\frac{e^{-\eta}}{\eta}d\eta + \frac{1}{2}\int_{\eta_b}^{\infty}\frac{e^{-\eta}}{\eta}d\eta, \tag{7.73}$$

where η_a indicates an η evaluated for a particular time at a radius of 1 m from the pumping well and η_b indicates η is evaluated at a radius 1 m from the pumping well, 1 year later. Since the values of η we are interested in here are undoubtedly small, we can expand the integrals in Equation 7.73 using the convergent series of Equation 7.71; combining the two series termwise yields the series expression for $\Delta\theta$:

$$\Delta\theta = \frac{1}{2}\left[\ln\left|\frac{\eta_a}{\eta_b}\right| + \sum_{k=1}^{\infty}\frac{(-1)^{k+1}(\eta_b^k - \eta_a^k)}{k \cdot k!}\right]. \tag{7.74}$$

In order to apply Equation 7.74 to the question of whether or not our system is at equilibrium, we will need actual numbers with which to calculate η_a and η_b. Since the time $t \approx 400$ years, the result won't be very sensitive to the exact values we choose for S, T, and so on. We will take $S = 1.0 \times 10^{-5}$ (a reasonable value for a confined aquifer) and $T = 1.0 \times 10^{-3}$ m^2/s (e.g., an aquifer thickness of 1000 m and a hydraulic conductivity of 1.0×10^{-6} m/s, which is a good conductivity for a sandy aquifer). To find the amount of drawdown over a year's time, we will also need to know the pumping rate; for this, we will use $Q = 1.0 \times 10^{-3}$ m^3/s, which is 1 L/s (or about 13 gpm in US customary units). Finally, 400 and 401 years translate into 1.262304×10^{10} and $1.26545976 \times 10^{10}$ seconds, respectively. Using these values, we can now define values for η_a and η_b:

$$\eta_a = \frac{1}{\frac{1.0\times10^{-3}}{1.0\times10^{-5}}1.262304 \times 10^{10}}, \tag{7.75}$$

$$\eta_b = \frac{1}{\frac{1.0\times10^{-3}}{1.0\times10^{-5}}1.26545976 \times 10^{10}}, \tag{7.76}$$

or $\eta_a = 7.92202 \times 10^{-13}$ and $\eta_b = 7.92207 \times 10^{-13}$. Plugging these values into Equation 7.74, we find $\Delta\theta \approx 1.248 \times 10^{-3}$ or, substituting for Q and T, we find a drawdown over this 1-year time period of about 0.016 m.

So, even after 400 years of pumping at a very modest rate, the aquifer still is not completely at equilibrium. In fact, given our scenario, the drawdown will always continue to increase, although ever more slowly over time. Other conditions, not allowed for in this model, may interfere with this conclusion.

For example, although the Plain of La Mancha of our imagination may by infinite, real aquifers are not unlimited in areal extent. If the aquifer is bounded by no-flow (or low-flow) boundaries, drawdown will accelerate as the cone of depression intersects the aquifer bounds. If the aquifer boundaries include constant head sources such as rivers or lakes, or if there is areally distributed recharge to the aquifer, the cone of depression may eventually stabilize and come into equilibrium with pumping. These situations are outside the situations considered by our model; however, the reader may wish to consider how the governing equation could be modified to take account of, for example, areal recharge.

7.8.2 The shape of time and space

It's worth stopping to ask about the "meaning" of the similarity variable $\eta = r^2/4Dt$. When I first saw a similarity solution to a partial differential equation demonstrated, I was confused. Why would a ratio of (squared) distance and time somehow be able to act together as a single variable? Most of the dimensionless variables we have looked at in the course of this text may seem a little odd, but they are really just scaled over some range. The similarity variable η is in an entirely different category and, as a result, it seems obscure and a little bit mysterious.

The truth is that I can't tell you why space and time scale the way they do in the similarity variable η, any more than we can say why ordinary space has three dimensions,[2] or why like charges repel and opposite charges attract. Probably, the best we can do is to say "because time and space are shaped that way."[3] Although I can't say *why* space and time scale together in the similarity variable, we *can* profitably ask ourselves what the scaling tells us about the way potential varies in our model.

The most interesting thing about the solution for η is that it tells us about what is happening in both time and space, simultaneously. That is, any plot of the solution can be looked at as either telling us what the potentiometric surface looks like as a function of distance from the pumping well at any particular time or, alternatively, how the potentiometric surface will behave at a specific radius away from the pumping well for all times, $0 < t < \infty$. This duality never ceases to amaze me, even after all of the years I have been contemplating this solution. To me, it is the mathematical equivalent of the well-known Rinzai Buddhist kōan: "Two hands clap and there is a sound. What is the sound of one hand?" (attributed to Hakuin Ekaku, 1686–1769).

To add to the mystery, there is the fact that, on a plot of θ versus η, time runs "backward" (from right to left, rather than from left to right) in comparison to distance. That is, because $\eta \propto r^2/t$, increasing distances from the pumping well run left to right as r goes from $0 \rightarrow \infty$. On the other hand, increasing time goes from right to left, with the initial time $t = 0$ infinitely far out on the left.

As $t \rightarrow \infty$, $\eta \rightarrow 0$, which accounts for the strong curvature of the plot as drawdown asymptotically goes to infinity near $\eta = 0$ (remember from Section 7.8.1 that drawdown never reaches steady state in the Theis model, but will continue to increase forever). Because of this odd depiction of temporal behavior, the plot of θ versus η goes asymptotically to the initial condition as $\eta \rightarrow \infty$, while the time compression as η nears the θ-axis causes the plot to launch off to infinity.

7.9 Violating the model assumptions

Of course we know that, in the real world, no aquifer is infinite in areal extent. As a result, sooner or later the cone of depression from our pumping well is going to intersect some kind of a boundary; when this happens, we will have violated our model assumptions and our model will no longer be a good representation of reality. When we violated the assumptions in our model of exponential decay (see Section 2.7), we found our model was no longer capable of fitting the observations. As a result, we had to extend our model before we could adequately fit the data.

What happens when we violate the assumptions of the Theis model? The consequences of violating the assumption of infinite areal extent are the easiest to evaluate. This assumption could be violated in two ways[4]: (i) our cone of depression could intersect a constant-head (recharge) boundary, or (ii) it could intersect a no-flow boundary.

The results of these two violations of the areally infinite aquifer assumption are intuitively obvious: if we intersect a recharge boundary (a river, a lake, etc.), the cone of depression will slow its expansion, and eventually the water level in the well will stabilize as the recharge from the boundary comes into equilibrium with the pumping rate. If, on the other hand, the cone of depression intersects a no-flow boundary (an impermeable region such as an unfractured granitic intrusion, a grout curtain, etc.), both drawdown in the well and the extent of the cone of depression will begin increasing more rapidly to make up for the lost yield from that part of the aquifer.

The effect of both of these situations on the data collected in either observation wells or the pumping well is immediately obvious to a trained observer. When we see these types of deviations from our model-based expectations, we know that one of the model assumptions has been violated, and we can often make a very good guess as to which assumption is in question. We can even make a good guess as to how far from the pumping well the boundary is by setting the drawdown in our model to some suitably small value (e.g., 0.1 m; representing the leading edge of the cone of depression) and calculating the radius at which that drawdown existed at the time we first noted the deviation from ideal behavior. Of course, this method won't tell us the *direction* in which the boundary lies, because the Theis equation assumes radial symmetry (all directions are equal), but a good understanding of the local geology and hydrology will often allow us to make an informed conjecture about this, too.

This brings up an important point about models that, until now, we have skirted around. As we have stated several times, models are a simplification—a simplified or an idealized representation of reality. Because they are *not* reality, *models are always wrong*. It has been said that "...all models are wrong; some models are useful" (George Box, 1919–2013). I prefer to say that "all models are wrong, but the way in which a model is wrong teaches us about the system." By this reasoning, whenever your model doesn't fit the observations, you should always ask yourself "why not?" If you can figure out why the model doesn't reproduce the data, you will have learned something important about the system; in that way, the time you spent developing a model that is "wrong" will still be time well spent.

7.10 Conclusions

In this chapter we examined a seminal contribution to groundwater modeling: the Theis equation, which is a mathematical representation of the impact of a pumping well on the potentiometric surface of an areally extensive aquifer. Although the governing equation is just the 1D transient diffusion equation in radial coordinates, the derivation is somewhat unusual because of the inclusion of a pumping well at $r = 0$. Furthermore, nondimensionalization of the governing equation illustrated a very interesting and powerful transform: the similarity variable, which incorporates two independent variables that scale together into a single variable that represents both.

In solving the Theis equation, we became familiar with an important function called the "exponential integral," known to hydrogeologists as the "well function," and we discussed the role of tabulated functions in applied mathematics. We also demonstrated that the Theis equation predicts that no steady state will ever be reached for the case of constant pumping in an areally infinite aquifer. In real life, however, we know that no aquifer is infinite in extent. Rather than a drawback, this is actually a benefit, and we discussed the fact that model deviations from field observations can actually be useful, telling us about characteristics of the system that are beyond the reach of our idealized models.

The Theis equation is a subtle and elegant model, and it has been found to be surprisingly robust to small violations of the model assumptions (i.e., small changes in aquifer thickness b or heterogeneous distributions of hydraulic conductivity). There is much to be learned from careful study of this representation of a common physical situation; hopefully, you will take this brief introduction as a challenge to learn more about this model, and apply your learning to your own situations of interest.

7.11 Problems

1. Using Equation 7.14 as your starting point, write a finite difference representation of the 1D transient diffusion equation in radial coordinates.

2. Construct a 2D finite difference model of the transient diffusion equation in Cartesian coordinates that includes the influence of a well, pumping at a constant rate. Compare the model output to the analytical solution from the Theis equation. How do your finite difference approximations compare with the analytical solution? HINT: Can you make use of symmetry when setting up your finite difference grid?

3. A semi-infinite solid (i.e., an infinite half-space) is initially at a constant temperature. At time $t = 0$, the temperature at $x = 0$ is changed from the initial temperature T_∞ to a new temperature T_0. Derive the governing equation and boundary conditions for the problem, nondimensionalize them, and solve. HINT: There is no "closed-form" solution to this problem; instead, the solution is an integral that is a tabulated function.

4. Derive the transient diffusion equation for 3D radial coordinates (r, θ, z).

5. Derive the transient diffusion equation for 3D spherical coordinates (r, θ, ϕ). This is considerably more difficult than the radial coordinates problem; you may want to seek advice from your instructor, or look in one of the standard texts on heat transfer or groundwater.

Notes

1 The use of "$\$$" for drawdown can be attributed to Professor D.K. Kreamer at the University of Nevada, Las Vegas. Commonly, "s" is used to indicate drawdown in hydrologic literature. However, Professor Kreamer, who was my MSc advisor, felt that it was too easy to confuse "s" (drawdown) with, for example, "S" (storativity) or "S" (saturation), and so on, and devised the $\$$-notation as an alternative.

2 Although our experience tells us there are 3 dimensions in ordinary space, some versions of string theory assert that space may have as many as 11 dimensions.

3 Unless you prefer René Descartes's ontological argument.

4 Of course, this assumption could be violated in an infinite number of ways that grade between the two described here. In this case, we are addressing the end-members of the problem for the sake of clarity.

References

Abramowitz, M. and Stegun, I. (1972) *Handbook of Mathematical Functions with Formulas, Graphs, and Tables*, Dover Publications, New York.

Theis, C.V. (1935) The relation between the lowering of the piezometric surface and the rate and duration of discharge of a well using ground-water storage. *Eos, Transactions of the American Geophysical Union*, **16**, 519–524.

CHAPTER 8

The transport equation

Chapter summary

Up to this point, we have concentrated on the problem of understanding potential flow—nominally, the flow of water in a porous medium, although the equations we have developed apply equally to species diffusion or the diffusion of heat. In this chapter, however, we will go beyond diffusion in a stationary medium and examine the movement of an intensive quantity that is being transported by a combination of diffusion and advection in a moving fluid. This conceptual model applies to the transport of heat by groundwater (e.g., in a hydrothermal system or in the vicinity of a buried package of high-level nuclear waste), a conservative species, such as lithium bromide, added to groundwater as a tracer, or any contaminant that moves in solution as part of an advecting fluid. The transport equation is even general enough to cover situations such as modeling the temperature distribution in a tectonic plate at a spreading center or a subduction zone; as such, it is the most general equation we will examine in this text, and will require that we bring together our understanding of many of the most important aspects of the previous chapters into a coherent whole.

8.1 The advection–dispersion equation

Our discussions of the Laplace equation, the Poisson equation, and the transient diffusion equation all share a common thread: in each of those three equations, the ultimate distribution of the dependent variable is driven by a process of diffusion. In essence, diffusion is a statistical-mechanical process in which random motions of molecules drive the value of the dependent variable from regions of high concentration to regions of lower concentration; this is also called *potential flow*. Take, for example, a bottle of perfume opened in a room of still air. The perfume molecules are at a high concentration in and near the bottle; by random (Brownian) motion, some of the molecules are likely to move away from the bottle and into the open air. Any given molecule is just as likely to move toward the bottle as away from the bottle; however, there are so many molecules

Models and Modeling: An Introduction for Earth and Environmental Scientists, First Edition. Jerry P. Fairley.
© 2017 John Wiley & Sons, Ltd. Published 2017 by John Wiley & Sons, Ltd.
Companion website: www.wiley.com/go/Fairley/Models

near the bottle that the overall tendency will be for the molecules to move away from the bottle and spread out into the room. If enough time is allowed to pass without disturbing the air in the room, the perfume molecules would eventually be evenly distributed throughout the room, and the situation would have reached steady state. This simple scenario provides a clear demonstration of the process of diffusion, in which molecules (or energy, in the case of the diffusion of heat or pressure) tend to move from regions of high potential to regions of lower potential, with the final (steady-state) concentration being defined by the overall characteristics of the system (the boundary conditions, the distribution of properties, etc.).

Note that, in the previous example, it was specified that the air in the room was still. If the air in the room is moving, the process becomes much more complex, and we can no longer describe either the transient or the steady-state distributions of molecules using one of the diffusion-based equations we have studied up to this point. Once the dependent variable is being influenced by both the process of diffusion and the movement of a carrier fluid (an "advecting" fluid), we are undertaking an investigation of a much deeper and more intricate problem. The equation that we use to model these transport processes is known either as the "advection–dispersion equation" (ADE), or the "convection–diffusion equation" (CDE); I personally prefer the former, although the terms are synonymous. The equation is also used to describe the time evolution of a particle's probability distribution function (pdf) in classical and quantum dynamics, in which case it is usually referred to as the "Fokker–Planck equation," after Adriaan Fokker (1887–1972) and Max Planck (1858–1947). In 1D Cartesian coordinates the ADE is written:

$$\frac{\partial}{\partial x}[U(x, t)C(x, t)] - \frac{\partial}{\partial x}\left[K\frac{\partial C(x, t)}{\partial x}\right] = \frac{\partial C(x, t)}{\partial t}, \tag{8.1}$$

where U is fluid velocity, C is concentration (or temperature, for heat transfer problems), K is conductivity (diffusivity for species transport), and x and t are the spatial and temporal coordinates, respectively. From left to right, the terms in Equation 8.1 are the advection term (a first derivative), the diffusion term (a second derivative) and, on the right-hand side of the equals sign, the temporal derivative.[1] In grad notation the ADE is written:

$$\nabla[UC] - \nabla[K\nabla C] = \frac{\partial C}{\partial t}. \tag{8.2}$$

The new factor in this equation is the advective term, which quantifies the movement of the fluid that carries the transported quantity along with it. The other two terms—the diffusion term and the temporal derivative—are essentially the same as those we have dealt with in previous chapters.

8.2 The problem child

The ADE is a much more difficult equation to work with than any of the diffusion equations (or the exponential decay equations) we have examined so far. Why would that be? There are several reasons. First of all, the ADE, as an equation, does not generally stand on its own. The dependent variable in Equation 8.1 is concentration (or temperature, or some other transported quantity), but note that the advective term,

$$\frac{\partial}{\partial x}\left[U(x,t)C(x,t)\right],\tag{8.3}$$

contains within the partial derivative a function that describes the velocity of the fluid transporting the dependent variable. This velocity can be a function of time and space, and it cannot be solved for in the context of the ADE itself. As a result, there must be a separate equation (or system of equations) that is solved to find the velocity field U (note that, in the general problem, U can be a function of three spatial coordinates and of time, although Equation 8.1 is only 1D transient). Usually, this means we begin by solving Laplace's equation, Poisson's equation, or the diffusion equation to find the flowfield of the carrier fluid, then we use the calculated flowfield as an input to the ADE. If the transported quantity is dilute (e.g., the species concentration is too small to change the fluid density, or the temperature differential is too small to induce convection due to buoyant forcing or coupled thermal-hydrologic gradients), this approach is adequate. If the effect of changes in the transported quantity over time or space is too great to be neglected, however, the solution of the ADE will "feedback" and perturb the solution of the diffusion equation, which in turn exerts a forcing on the ADE, which feeds back on the diffusion equation, and so on. As a result, the ADE and the driving diffusion equation cannot usually be solved in sequence; rather, we are required to solve the coupled equations simultaneously—a much more difficult proposition than solving them consecutively.

   Although the coupled solution of the fluid flow and transport equations is seldom analytically tractable, it is still possible to solve the equations using numerical approximation. The flow and transport equations are amenable to solution by finite differences, integrated finite differences (see, e.g., Patankar (1980)), or finite element methods (e.g., Huyakorn and Pinder (1983) and Istok (1989)). However, even these numerical methods are not completely without difficulty. Especially in situations where the diffusion term is small relative to the advective term, numerical schemes can suffer from large amounts of *numerical dispersion* (i.e., nonphysical smearing of otherwise sharp fronts and erroneously high or low apparent diffusion coefficients), and may require excessively small timesteps or extremely high-resolution gridding to avoid numerical instability and obtain an acceptable solution. Finite difference methods are especially prone to problems with numerical dispersion; finite element methods, on the other hand, are less likely to suffer from numerical dispersion, but may encounter difficulties with numerical instability.

We will return to these difficulties with the ADE later in this chapter, when we have a better understanding of the behavior implied by the equation and some of its analytical and numerical characteristics. In the meanwhile, we will glance back at a problem we examined earlier, which we unknowingly left incomplete.

8.3 The Augean Stables, revisited

You may remember that, in Chapter 6, we told the story of the Twelve Labors of Hercules. Of special interest to us was the cleaning of the Augean Stables—the labor in which Hercules, under orders from King Eurystheus, cleaned out the stables of King Augeus. Although the stables hadn't been cleaned in over 30 years, Hercules was able to clean them in a single day by diverting the River Alpheus up onto the floodplain upon which the stables were located, washing out the accumulated debris. In our rendering of the story, the diversion of the river caused an instantaneous rise in head at the boundary of a high-permeability sandstone aquifer, and this sudden change in head propagated through the aquifer in a fashion that we modeled using the transient diffusion equation.

Although we didn't discuss this in Chapter 6, it is a little-known fact that King Augeus, whose palace was on the hill between the River Alpheus and the River Peneus (the two boundaries of our model domain), obtained all his water from a well, located on the hill and completed in the sandstone aquifer halfway between the two rivers (see Figure 8.1). King Augeus thought he got a pretty good deal when Hercules cleaned out his stables, but did he really? Let's find out.

8.4 Defining the problem

We know that nitrates are extremely soluble in water, and high nitrate concentrations in drinking water have been linked to a number of health problems in

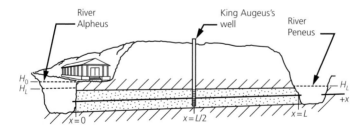

Figure 8.1 Schematic diagram of the Augean Stables problem, revisited. In this revised problem, King Augeus's well is located on the hill between the two model boundaries at $x = L/2$.

humans. For this reason, the United States Environmental Protection Agency (EPA) has set a "maximum contaminant level" (MCL) of 10 mg/l in drinking water. Nitrates can come from many sources (e.g., from geologic deposits of nitrate salts, runoff from agricultural areas, etc.), and stables, barns, feedlots, and other facilities associated with livestock are definitely a known source. As a result, we can assume the water from the River Alpheus contains an elevated concentration of nitrates after running through the Augean Stables, and that this contaminated water will find its way into the aquifer. The question that concerns us, therefore, is as follows: "will the concentration of nitrates in the water of King Augeus's well exceed the maximum concentration allowable for human consumption? If so, in what timeframe?" Within the context of our problem-solving framework, we have the following:

FIND: The time-varying concentration of nitrates in the aquifer between the Rivers Alpheus and Peneus; in particular, we want to know the concentration of nitrates as a function of time in King Augeus's well.

KNOWN: Most of what we know about the problem is stated in the description of Chapter 6. In addition, we can add here the following:

- We "know" the concentration of nitrates at the boundary formed by the River Alpheus (the concentrations are at least knowable—they could be measured if we were on-site).
- Similarly, we "know" the background levels of nitrate in the aquifer (which we could presumably find from sampling the well) and in the River Peneus.
- The flux of water in the aquifer, as a function of time and position, is known from Equation 6.106.

ASSUME: We will make the following assumptions, in addition to the assumptions previously stated in Chapter 6:

- King Augeus's well is not being pumped (we can assume that dipping a bucket in the well every now and again doesn't disturb the flow field of the aquifer).
- The nitrate concentrations are not affected by precipitation or dissolution reactions, and there is no sorption, retardation, or any decomposition reactions.
- Changes in density of the fluid due to changing concentrations of nitrate are negligible.
- We will assume the concentration of nitrates at the River Alpheus remains constant for the duration of interest.
- We will assume the initial concentration of nitrates in the aquifer and the concentration of nitrates in the River Peneus are the same.

APPROACH: Using mass balance, we will construct the governing equation for the system. The solution to Equation 6.39, subject to the conditions in Equations 6.40–6.42, will be used to develop the required expression for fluid velocity ($U(x, t)$ in Eq. 8.1). We will attempt to solve the governing equation analytically; if that is not practical, we will attempt an approximate numerical solution (or, at least, describe a method for achieving a numerical solution).

Although we may need additional assumptions as we go along (and we will, of course, keep track of those additional assumptions), we can list these as we need them. In the meanwhile, we will attempt to carry out our approach and see where it leads us.

8.5 The governing equation

We begin by defining a CV; in this case, the CV is the same as that defined for the Augean Stables problem in Chapter 6 (see Figure 6.3), although the fluxes will be different, as will be seen later. The CV for this problem, with the attendant fluxes labeled, is shown in Figure 8.2.

As always, the statement of mass balance is given by

$$M_{in} = M_{out} + \Delta M. \tag{8.4}$$

The steps we take from here should be familiar: we will start by expanding the terms in Equation 8.4 on the basis of our understanding of the problem as portrayed in Figure 8.2. Although the steps we take in the expansions should be familiar, this time we have a few extra fluxes, and these will deserve some additional comment. We can begin with the M_{in} term. We can see in Figure 8.2 that we have two fluxes in. In addition to the diffusive flux q_x, we have the advective flux $C_x U_x$, which is the flux of nitrates (at concentration C_x) being carried in by the advecting fluid (at a velocity U_x). This flux will need to be multiplied by the area of the CV through which the flux is passing, the fluid density, and so on. To make things easy on ourselves, we will make the nitrate

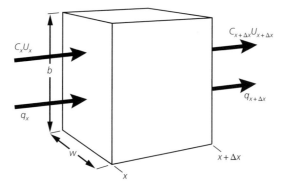

Figure 8.2 Control volume for the Augean Stables redux problem. In this case, the q_x and $q_{x+\Delta x}$ fluxes refer to the diffusive flux of nitrates, rather than the water flux, as they did in Figure 6.3. In the figure (and in the text), U is the water velocity.

concentrations in mg/kg, rather than mg/l (these are very close, but there would be a small correction for the density of water at the temperature of the aquifer).

The other flux going into the CV is q_x, which is the flux of nitrates resulting from diffusion of the nitrates in the water. If we put these two fluxes together, we have a statement of the mass fluxes coming into the CV,

$$M_{in} = C_x U_x wb\rho \Delta t + q_x wb\rho \Delta t, \tag{8.5}$$

where ρ is the water density. Similarly, the M_{out} term is

$$M_{out} = C_{x+\Delta x} U_{x+\Delta x} wb\rho \Delta t + q_{x+\Delta x} wb\rho \Delta t. \tag{8.6}$$

The final term, ΔM, is

$$\Delta M = \Delta C wb \Delta x \rho. \tag{8.7}$$

Inspection of these terms shows they all have units of mass [M] (with concentrations in mg/kg, the mass is given in units of mg of nitrate), as should be the case. Substituting the three terms into Equation 8.4, we get

$$C_x U_x wb\rho \Delta t + q_x wb\rho \Delta t$$
$$= C_{x+\Delta x} U_{x+\Delta x} wb\rho \Delta t + q_{x+\Delta x} wb\rho \Delta t + \Delta C wb\rho \Delta x. \tag{8.8}$$

We can rearrange Equation 8.8 to get like terms together; at the same time, we will cancel the $wb\rho$ factors that occur in all the terms. The result is

$$(C_{x+\Delta x} U_{x+\Delta x} - C_x U_x) \Delta t + (q_{x+\Delta x} - q_x) \Delta t = -\Delta C \Delta x. \tag{8.9}$$

Next, we divide through by $\Delta x \Delta t$ to find

$$\frac{C_{x+\Delta x} U_{x+\Delta x} - C_x U_x}{\Delta x} + \frac{q_{x+\Delta x} - q_x}{\Delta x} = -\frac{\Delta C}{\Delta t}. \tag{8.10}$$

Taking the limit as $\Delta x, \Delta t \to 0$, we arrive at the differential form of Equation 8.10, which is

$$\frac{\partial}{\partial x}[CU] + \frac{\partial q}{\partial x} = -\frac{\partial C}{\partial t}. \tag{8.11}$$

These steps should all be familiar, notwithstanding the addition of the advective term. When we reached this stage in previous chapters, our next step would be to substitute Darcy's law for q. In the present situation, we can't substitute Darcy's law, because it applies to the flow of fluid in a porous medium, whereas here we are talking about the diffusion of chemical species (nitrate) in a fluid. Although we can't use Darcy's law, we are completely justified in using an essentially identical law that governs the diffusion of species called "Fick's law",

$$q = -D_{AB}\frac{dC}{dx}, \tag{8.12}$$

where D_{AB} is the diffusivity of species A in substance B (in this case, the diffusivity of nitrate in water; [L²/T]). Substituting Equation 8.12 into Equation 8.11, we have

$$\frac{\partial}{\partial x}[CU] - \frac{\partial}{\partial x}\left[D_{AB}\frac{\partial C}{\partial x}\right] = -\frac{\partial C}{\partial t}. \tag{8.13}$$

In Equation 8.13, we have brought the minus sign from Fick's law out from under the differential. The parameter D_{AB} (the diffusion coefficient) is considered to be a constant for our problem.[2] As a result, we can pull D_{AB} out from under the derivative to obtain

$$\frac{\partial}{\partial x}[CU] - D_{AB}\frac{\partial^2 C}{\partial x^2} = -\frac{\partial C}{\partial t}. \tag{8.14}$$

Equation 8.14 is the governing equation describing the transport of nitrate in the Augean Stables aquifer (as was discussed earlier, the flow field U is described by the solution to the transient diffusion equation). To make this equation specific to our problem, we need only prescribe boundary and initial conditions. On the basis of our assumptions and discussion in Section 8.4, we can state the auxiliary conditions as

$$C(x, t = 0) = C_L, \tag{8.15}$$

$$C(x = 0, t) = C_0, \tag{8.16}$$

$$C(x = L, t) = C_L. \tag{8.17}$$

With these conditions, we have specified that the initial concentration of nitrates in the aquifer, as well as the fixed concentration of nitrates at the boundary $x = L$, are the same and equal to C_L. Furthermore, the concentration of nitrates at the boundary $x = 0$ is fixed for all times $t \geq 0$ at the constant value C_0. This value C_0 is known to us—or, at least, it could in principle be measured in the field.

8.6 Nondimensionalization

We are now at a familiar juncture: we have a (dimensional) governing equation and a set of auxiliary conditions. From past experience, we know the next step is to nondimensionalize the equation and conditions. Because this problem will present some new features in addition to those we have come to expect, we will address the various parts of the nondimensionalization separately, then bring them together after we have had the opportunity to discuss each aspect in turn.

8.6.1 The dependent variable

As usual, we will begin by normalizing the dependent variable, C. From inspection of the boundary and initial conditions, we can see that the concentration of nitrates is likely to vary over a range of $C_L \leq C \leq C_0$. If we assume the initial concentration of nitrates in the aquifer, C_L, is equal to zero, we can use the normalization:

$$\theta_N = \frac{C(x, t)}{C_0}. \tag{8.18}$$

We will use the subscript "N" to indicate this particular θ refers to nitrate concentration. In this case, it will be necessary to employ the subscript for the sake of clarity, because we will use the results from Chapter 6, in which θ indicated dimensionless head.

This normalization scheme maps the concentration C onto a range $0 \le \theta_N \le 1$. Alternatively, we could have used the nondimensionalization scheme,

$$\theta_N = \frac{C(x, t) - C_L}{C_0 - C_L}, \tag{8.19}$$

which is our usual method of normalizing our dependent variable, and would allow us to include a background concentration of nitrates, rather than having to assume a zero background concentration. For low background concentrations, the two normalization schemes ultimately give equivalent answers, but in this case the second method (i.e., Eq. 8.19) gives a somewhat more complicated form to our nondimensional governing equation.[3] Solving Equation 8.18 for $C(x, t)$ and taking the first and second derivatives, we have

$$C(x, t) = C_0 \theta_N, \tag{8.20}$$

$$dC = C_0 d\theta_N, \tag{8.21}$$

$$d^2 C = C_0 d^2 \theta_N. \tag{8.22}$$

8.6.2 The independent variables

Because we want to use the results from Chapter 6 in the present problem, we will have to use the same nondimensionalization scheme here for any of the variables shared by the two problems. We can therefore normalize x to the scale of the domain, L, as we did in Chapter 6, by letting

$$\xi = \frac{x}{L}. \tag{8.23}$$

We then find that

$$x = L\xi, \tag{8.24}$$

$$x^2 = L^2 \xi^2, \tag{8.25}$$

$$dx = L d\xi, \tag{8.26}$$

$$dx^2 = L^2 d\xi^2. \tag{8.27}$$

For t there is, as usual, no obvious range upon which to normalize. As a result, we define an arbitrary characteristic time t_c and proceed as:

$$\tau = \frac{t}{t_c}, \tag{8.28}$$

$$t = t_c \tau, \tag{8.29}$$

$$dt = t_c d\tau. \tag{8.30}$$

At this point, these normalizations should all be familiar to you. We will, of course, define t_c in the process of substituting the non-dimensional variables into the governing equation; in the meanwhile, we can continue on to the remaining quantity in need of attention.

8.6.3 The velocity

The final thing we have to work out is how to non-dimensionalize the velocity function $U(x, t)$. It is often more difficult to work out the way in which functions should be non-dimensionalized when they appear in the governing equation, because some characteristic quantity must be identified that is both a constant and is physically meaningful. In the present case, we can work backwards, starting with the dimensionless flux v:

$$v = -\frac{d\theta}{d\xi}, \tag{8.31}$$

$$v = 1 + 2 \sum_{n=1}^{\infty} e^{-(n\pi)^2 \tau} \cos(n\pi \xi). \tag{8.32}$$

Equations 8.31 and 8.32 are the same as Equations 6.105 and 6.106, and are repeated here for convenience of reference. We also know from Equation 6.104 that:

$$v = \frac{q}{q_c} = \frac{q}{-K \frac{(H_L - H_0)}{L}}. \tag{8.33}$$

Finally, the relationship between flux and the actual velocity of the fluid is:

$$U = \frac{q}{\phi}, \tag{8.34}$$

where ϕ is porosity. Applying these relationships, we can see the fluid velocity function U is equal to:

$$U = U_c v, \tag{8.35}$$

where:

$$U_c = -\frac{K(H_L - H_0)}{\phi L}. \tag{8.36}$$

8.6.4 Making the substitutions

On the basis of our work in the previous sections, we can now nondimensionalize the governing equation and auxiliary conditions. Substituting Equations 8.20–8.22, 8.26, 8.27, 8.30, and 8.35 into the governing equation (Eq. 8.14), we obtain

$$\frac{C_0 U_c}{L}\frac{\partial}{\partial \xi}[v\theta_N] - \frac{C_0 D_{AB}}{L^2}\frac{\partial^2 \theta_N}{\partial \xi^2} = -\frac{C_0}{t_c}\frac{\partial \theta_N}{\partial \tau}. \tag{8.37}$$

To simplify, we can cancel the C_0 factors, multiply both sides by L^2, and divide both sides by -1 and D_{AB}. The result is

$$\frac{\partial^2 \theta_N}{\partial \xi^2} - \frac{U_c L}{D_{AB}}\frac{\partial}{\partial \xi}[v\theta_N] = \frac{L^2}{D_{AB}t_c}\frac{\partial \theta_N}{\partial \tau}. \tag{8.38}$$

We can recognize L^2/D_{AB} as a characteristic time constant; therefore, we would like to set $t_c = L^2/D_{AB}$. Unfortunately, we don't have the freedom to do so in this case. If we look back at Equation 8.32, we can see that it already contains a dimensionless time τ that is normalized on the characteristic time for the original Augean Stables problem. If we want to use the results from that analysis in the present problem, we will have to use the same dimensionless time. We will therefore set the characteristic time in Equation 8.38 to the value defined by Equation 6.38,

$$t_c = \frac{L^2 S}{T} = \frac{L^2}{D}, \tag{8.39}$$

where S and T are the aquifer storativity and transmissivity, respectively, and D is the hydraulic diffusivity, $D = T/S$. This is, of course, different from the species diffusivity (the coefficient of diffusion), D_{AB}, in Equation 8.38. When we substitute this characteristic time into Equation 8.38, the length scales (L) cancel and we are left with the ratio of the diffusivities. The units of all diffusivities are $[L^2/T]$; therefore, we can define a dimensionless parameter α,

$$\alpha = \frac{D}{D_{AB}}, \tag{8.40}$$

which, when we rewrite Equation 8.38, gives us

$$\frac{\partial^2 \theta_N}{\partial \xi^2} - \frac{U_c L}{D_{AB}}\frac{\partial}{\partial \xi}[v\theta_N] = \alpha\frac{\partial \theta_N}{\partial \tau}. \tag{8.41}$$

The remaining parameter in Equation 8.41, $U_c L/D_{AB}$, is perhaps unfamiliar to some readers. However, it is dimensionless (the units of U_c are those of velocity $[L/T]$, and the diffusion coefficient has units of $[L^2/T]$), and there are no undefined characteristic quantities that we can set to annihilate the dimensional parameters (i.e., to reduce their ratio to 1). As a result, we will define a new dimensionless number, λ:

$$\lambda = \frac{U_c L}{D_{AB}}. \tag{8.42}$$

The parameter we have called λ is widely known as the *Peclet number* (pronounced "peck-LAY"); it is often abbreviated using the symbol "Pe" or similar. The Peclet number quantifies the relative importance of advection to diffusion. Large Peclet numbers ($\lambda \gg 1$) indicate that advection is dominating over the process of diffusion; conversely, small Peclet numbers ($\lambda \ll 1$) imply that advection is relatively unimportant, and diffusion is the primary driving process for transporting the intensive quantity. Of course, if the Peclet number is on the order of 1, diffusion and advection are approximately equal in importance.

Our final statement of the dimensionless governing equation is

$$\frac{\partial^2 \theta_N}{\partial \xi^2} - \lambda \frac{\partial}{\partial \xi} [v\theta_N] = \alpha \frac{\partial \theta_N}{\partial \tau}, \tag{8.43}$$

with auxiliary conditions:

$$\theta_N(\xi, \tau = 0) = 0, \tag{8.44}$$

$$\theta_N(\xi = 0, \tau) = 1, \tag{8.45}$$

$$\theta_N(\xi = 1, \tau) = 0. \tag{8.46}$$

In the following sections, we will discuss in more detail the impacts of Peclet numbers of various orders of magnitude on our problem. With our understanding of the Peclet number and the tools for solving PDEs that we have developed in previous chapters, we will consider the conditions under which we can find different types of solutions to our governing equation and auxiliary conditions.

8.7 Analytical solutions

Of course, we would like to be able to solve Equation 8.43 analytically. It may be, however, that in this case we can't immediately figure out *how* to solve it analytically or, if we can solve it, perhaps in the future we will run across an equation that we *cannot* solve analytically. Should we, then, give up, and proceed immediately to a numerical solution? In his book on learning to solve problems, George Polya describes the following dialog with an imaginary student who is trying to devise an approach to a new problem (Polya, 1973):

> You do not seem to find the problem too easy. *If you cannot solve the proposed problem, try to solve first some related problem.* Could you satisfy a *part of the condition?...Keep only a part of the condition, drop the other part.* What part of the condition is easy to satisfy? [emphasis in the original]

We will do as Professor Polya suggests, and examine individual parts of the problem that we *can* solve; in this way, we will (hopefully) obtain insights that will allow us to understand the entire problem.

8.7.1 Steady-state solutions

If we are looking for a way to consider only part of the problem, the first thing we can do is set the temporal derivative, $\partial \theta_N / \partial \tau$, equal to zero. This gives us the late-time behavior of the system; that is, the behavior of the system after it has reached steady state. The steady problem is described by an ODE in ξ,

$$\frac{d^2 \theta_N}{d\xi^2} - \lambda \frac{d}{d\xi} [v \theta_N] = 0. \tag{8.47}$$

This give us three possible cases we can examine: $\lambda \ll 1$, $\lambda \gg 1$, and $\lambda \approx 1$.

8.7.1.1 Diffusion dominates

For the case in which the Peclet number, λ, is much less than one, diffusion dominates, and advection is negligible in comparison.[4] In this case, we can set the advective derivative to zero, and we have

$$\frac{d^2 \theta_N}{d\xi^2} = 0. \tag{8.48}$$

Recognizing this equation as identical to the "Elysian Fields" equation of Chapter 3, we can immediately write the solution as

$$\theta_N = 1 - \xi. \tag{8.49}$$

8.7.1.2 Advection dominates

If the Peclet number should happen to be numerically much greater than 1, advection dominates. In this situation, the influence of diffusion is small and can be neglected; we set the diffusion term to zero and, dividing through by $-\lambda$, we obtain

$$\frac{d}{d\xi} [v \theta_N] = 0. \tag{8.50}$$

We can expand the derivative using the product rule

$$\theta_N \frac{dv}{d\xi} + v \frac{d\theta_N}{d\xi} = 0. \tag{8.51}$$

The derivative of v with respect to ξ is

$$\frac{dv}{d\xi} = -2\pi \sum_{n=1}^{\infty} n e^{-(n\pi)^2 \tau} \sin(n\pi \xi), \tag{8.52}$$

which can be substituted into Equation 8.51. With a little rearranging, we get

$$\frac{d\theta_N}{d\xi} = \theta_N \frac{dv/d\xi}{v},$$
(8.53)

where $dv/d\xi$ is given by Equation 8.52, and v is

$$v = 1 + 2\sum_{n=0}^{\infty} e^{-(n\pi)^2\tau}\cos(n\pi\xi).$$
(8.54)

If we attempt to integrate both sides of Equation 8.53, we get a really horrific-looking equation

$$\int \frac{1}{\theta_N}\frac{d\theta_N}{d\xi}\,d\xi = 2\pi\int \frac{\sum_{n=0}^{\infty} ne^{-(n\pi)^2\tau}\sin(n\pi\xi)}{1 + 2\sum_{n=0}^{\infty} e^{-(n\pi)^2\tau}\cos(n\pi\xi)}\,d\xi.$$
(8.55)

However, since we are looking at the steady-state problem, we can be relatively certain that τ will be large, and for large τ the sums on the right-hand side of Equation 8.55 both go to zero. As a result, the integral on the right-hand side of Equation 8.55 $\rightarrow 0/1 = 0$, and our problem reduces to a much more tractable equation[5],

$$\int \frac{d\theta_N}{\theta_N} = A,$$
(8.56)

where A is the constant of integration from the integral of the right-hand side. Integrating the left-hand side, raising both sides to a power of e, and combining the constants of integration yields the solution:

$$\theta_N = A.$$
(8.57)

When we set the temporal derivative to zero, we no longer needed the initial condition given by Equation 8.44. Similarly, when the problem was reduced to a first-degree ODE (Eq. 8.50), we can no longer have two boundary conditions or our problem will be overdetermined. Instead, we choose the boundary condition at $\xi = 0$ (Eq. 8.45); thus, $A = 1$, and we have for our solution

$$\theta_N = 1.$$
(8.58)

Although the solution to the advection-only problem given by Equation 8.58 may seem "trivial," we have learned a great deal about the form that any (steady-state) solution to our problem must take. The cases of $\lambda \ll 1$ and $\lambda \gg 1$ form the two extreme cases; any solution $\lambda \approx 1$ must lie between these two situations. The extreme cases, and the possible solution space between that encompasses all combinations of advection and diffusion, are plotted in Figure 8.3.

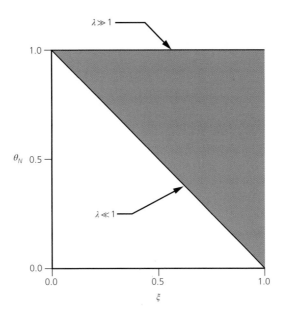

Figure 8.3 Steady-state solution space for the transport of nitrate in the Augean Stables aquifer. The line $\lambda \ll 1$ shows the steady solution for the diffusion-only problem, while the line $\lambda \gg 1$ is the steady solution for the case in which advection dominates over diffusion. All possible steady-state solutions must fall between these limits; therefore, the gray area denotes the space of possible steady-state solutions for this problem.

8.7.2 Transient solutions

When we set the temporal derivative equal to zero, we obtained an ordinary differential equation in ξ. This was an important simplification of the original problem, and made it relatively easy to find solutions. If we retain the temporal derivative, all possible simplifications of the original problem will be PDEs. We will still be able to find solutions, but they will be more complex than those given in Section 8.7.1.

The three cases we may consider are $\lambda \ll 1$, $\lambda \gg 1$, and $\lambda \approx 1$. Because the third of these cases ($\lambda \approx 1$) is really the solution of the full problem, we will discuss here the first two cases, and the ramifications of these partial solutions for the third case.

8.7.2.1 Diffusion dominates

For the case in which diffusion dominates, $\lambda \ll 1$. As a result, we can set the advection term to zero; the governing equation (Eq. 8.43) then becomes

$$\frac{\partial^2 \theta_N}{\partial \xi^2} = \alpha \frac{\partial \theta_N}{\partial \tau}. \tag{8.59}$$

For this equation, we will still need all three auxiliary conditions given in Equations 8.44–8.46 to find a complete solution.

It is true that the transient solutions to the governing equation are more difficult to find than the solutions to the corresponding steady-state problems. In this case, however, we can recognize that Equation 8.59 is almost identical to the original Augean Stables problem (Chapter 6), with the exception of the α parameter. If we apply the separation of variables method to Equation 8.59, we find the solution to be

$$\theta_N = 1 - \xi - \frac{2}{\pi} \sum_{n=1}^{\infty} e^{\frac{-(n\pi)^2 \tau}{\alpha}} \frac{\sin(n\pi\xi)}{n}, \tag{8.60}$$

which is the same as Equation 6.95, except that θ_N has been substituted for θ, and the factor of α shows up in the exponent of e, where it scales the dimensionless time to account for the differences in temporal scale between the original Augean Stables problem and the transport problem examined in this chapter. The solution to Equation 8.60 is plotted in Figure 8.4. Note that the late-time behavior of this solution converges on the linear, steady-state solution we identified in Section 8.7.1.1.

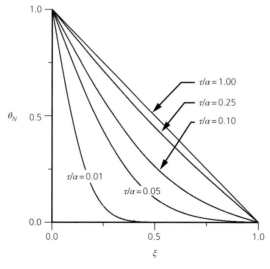

Figure 8.4 Transient solution of the diffusion-dominant (advection negligible) equation for nitrate transport in the Augean Stables aquifer. The concentration of nitrate can be seen sweeping through the aquifer from left to right. The approximate final (steady-state) concentration of nitrate in the aquifer differs from the line $\tau/\alpha = 1$ by an amount that is not visible at the scale of this figure (compare with the diffusion limit in Figure 8.3). The curves in the figure are the same as those in Figure 6.4, with dimensionless time scaled by the factor α.

8.7.2.2 Advection dominates

In cases where the Peclet number, λ, is much greater than 1, advection dominates, and diffusion can be neglected. In this situation, we can set the diffusion term equal to zero, and the governing equation becomes

$$\lambda \frac{\partial}{\partial \xi} [v \theta_N] = -\alpha \frac{\partial \theta_N}{\partial \tau}. \tag{8.61}$$

Of course, if λ is large, both the diffusion term and the temporal derivative may be relatively small by comparison, depending on the size of α. However, if we set both terms to zero, we obtain the steady-state advection dominated solution from Section 8.7.1.2. As a result, we will assume that λ and α are of the same order of magnitude, and retain the temporal derivative. The ratio of the two dimensionless paramters, λ (the Peclet number) and α (the ratio of characteristic times, or diffusivities), will therefore scale the relationship between the advection term and the temporal derivative.

Although Equation 8.61 is somewhat less difficult to solve than the full, transient ADE, it is still a formidable undertaking. When we expand the advective term (as we did in Eq. 8.51), we obtain

$$\frac{\partial v}{\partial \xi} \theta_N + v \frac{\partial \theta_N}{\partial \xi} = -\frac{\alpha}{\lambda} \frac{\partial \theta_N}{\partial \tau}. \tag{8.62}$$

The analytical expressions for v and $dv/d\xi$ are given by Equations 8.54 and 8.52 (repeated here for convenience):

$$v = 1 + 2 \sum_{n=1}^{\infty} e^{-(n\pi)^2 \tau} \cos(n\pi \xi), \tag{8.63}$$

and

$$\frac{dv}{d\xi} = -2\pi \sum_{n=1}^{\infty} n e^{-(n\pi)^2 \tau} \sin(n\pi \xi). \tag{8.64}$$

Equation 8.61 is a first-order PDE with variable coefficients (i.e., v and $dv/d\xi$ are not constants, but are rather functions of both time and space, as can be seen from Eqs. 8.63 and 8.64). In general, differential equations with variable coefficients are more difficult to solve than those with constant coefficients. Because of the complexity of finding an analytical solution to Equation 8.61, we will not undertake it here. However, intuition tells us that head in the aquifer will come to equilibrium more rapidly than the concentration of nitrate being transported through the aquifer. If we neglect the early time velocity changes, we have a steady-state flow field; v would be equal to one, and $dv/d\xi$ would be zero. Substituting a modified Peclet number, $\Lambda = \lambda/\alpha$, for the ratio of the two constants, the governing equation reduces to

$$\Lambda \frac{\partial \theta_N}{\partial \xi} + \frac{\partial \theta_N}{\partial \tau} = 0, \tag{8.65}$$

with auxiliary conditions:

$$\theta_N(\xi, \tau = 0) = 0, \tag{8.66}$$

$$\theta_N(\xi = 0, \tau) = 1. \tag{8.67}$$

Equation 8.65 is usually called the "advection equation," and it describes the displacement of species (or heat energy, etc.) by the process of advection. None of the methods that we have applied to the solution of PDEs or ODEs (direct integration, separation of variables) will be effective for solving the advection equation; instead, we will use the well-known, but unintuitive, *method of characteristics*, or "MOC" for short.

The MOC is a general technique for approaching first-order PDEs. Since it is usually applied only to first-order equations (although it can be extended to some higher order problems), it may seem somewhat restricted in its applicability; however, it can be used to find solutions to linear, quasi-linear, and fully non-linear equations. This versatility is quite uncommon among methods for solving differential equations, and makes it well worth learning despite its somewhat confusing nature. The essence of the method is to find a transformation that treats the independent variables in the governing equation as parametric variables, which are themselves dependent on a single variable. Once this transformation is found, it is used to break the governing PDE into a system of ODEs that are called the *characteristic equations* of the governing equation. The solution of these characteristic equations yields a set of "integral curves," also known as *characteristic curves* or, more simply, *characteristics*. These characteristics provide the solution to the governing equation.

If this all sounds a little fuzzy, an example solution of Equation 8.65 may clarify things somewhat (or at least shed some light on the behavior of the advection equation). To begin, we are looking for a transform that can make the independent variables ξ and τ in Equation 8.65 each a function of a single variable η,

$$\frac{d}{d\eta}\left[\theta_N\left(\xi(\eta), \tau(\eta)\right)\right] = F\left(\theta_N, \xi(\eta), \tau(\eta)\right). \tag{8.68}$$

To find this transform, we use the chain rule to expand the first derivative of θ_N with respect to η

$$\frac{d\theta_N}{d\eta} = \frac{\partial\theta_N}{\partial\xi}\frac{d\xi}{d\eta} + \frac{\partial\theta_N}{\partial\tau}\frac{d\tau}{d\eta}. \tag{8.69}$$

Comparing Equation 8.69 with Equation 8.65, we can see that

$$\frac{d\xi}{d\eta} = \Lambda, \tag{8.70}$$

$$\frac{d\tau}{d\eta} = 1. \tag{8.71}$$

We can therefore rewrite Equation 8.69 as

$$\frac{d\theta_N}{d\eta} = \Lambda \frac{\partial\theta_N}{\partial\xi} + \frac{\partial\theta_N}{\partial\tau} = 0. \tag{8.72}$$

We now have the following characteristic equations:

$$\frac{d\tau}{d\eta} = 1, \tag{8.73}$$

$$\frac{d\xi}{d\eta} = \Lambda, \tag{8.74}$$

$$\frac{d\theta_N}{d\eta} = 0. \tag{8.75}$$

These equations can all be solved by direct integration. Integrating both sides of Equation 8.73, we have

$$\int \frac{d\tau}{d\eta} \, d\eta = \int \, d\eta, \tag{8.76}$$

$$\int d\tau = \int \, d\eta, \tag{8.77}$$

$$\tau = \eta + c_1. \tag{8.78}$$

Similarly, we can integrate both sides of Equation 8.74 with respect to η

$$\int \frac{d\xi}{d\eta} \, d\eta = \Lambda \int \, d\eta, \tag{8.79}$$

$$\int d\xi = \Lambda \int \, d\eta, \tag{8.80}$$

$$\xi = \Lambda\eta + c_2. \tag{8.81}$$

Finally, integrating Equation 8.75 gives

$$\int \frac{d\theta_N}{d\eta} \, d\eta = 0 \int \, d\eta, \tag{8.82}$$

$$\int d\theta_N = 0 + c_3, \tag{8.83}$$

$$\theta_N = c_3. \tag{8.84}$$

To define the arbitrary constants, we need some conditions on the equations. For τ, we can require that all characteristic lines start from $\tau = 0$; this gives us the condition $\tau(\eta = 0) = 0$. We therefore have

$$\tau(\eta = 0) = 0 + c_1 = 0, \tag{8.85}$$

or $c_1 = 0$. For Equation 8.81, we will require each characteristic to start at $\eta = 0$ from some initial $\xi = \xi_0$ (i.e., a starting position ξ_0, unique to any given characteristic). Since we know from Equations 8.78 and 8.85 that $\eta = \tau$, we can require that any given characteristic must originate from its own definite initial value of ξ at an initial time $\tau = 0$:

$$\xi(\tau = 0) = (\Lambda)\,0 + c_2 = \xi_0, \tag{8.86}$$

which gives us $c_2 = \xi_0$. Substituting τ for η and ξ_0 for c_2 in Equation 8.81, we therefore have the following relationship between ξ and τ for any characteristic:

$$\xi - \Lambda\tau = \xi_0. \tag{8.87}$$

Because θ_N is a constant along any characteristic curve, Equation 8.87 tells us that the value of θ_N for a characteristic curve starting from $\xi = \xi_0$ at $\tau = 0$ translates with increasing values of τ to new values of ξ, so that the value of the left-hand side of Equation 8.87 is always equal to the constant ξ_0 (Figure 8.5). Another way of looking at this is to say that the initial value of θ_N, $\theta_N(\xi = \xi_0, \tau = 0)$ translates to new locations (i.e., new values of ξ), according to the relationship:

$$\xi = \Lambda\tau + \xi_0. \tag{8.88}$$

The value of θ_N at any initial ξ_0 is arbitrary (see Eq. 8.84), although it is constant along the characteristic emanating from that point. We are therefore free to define the initial values of θ_N across the domain to replicate the initial conditions of the problem. In this case, the function θ_N at time $\tau = 0$ is described by a Heaviside step function, equal to 1 at $\xi = 0$ and equal to zero for all $\xi > 0$.

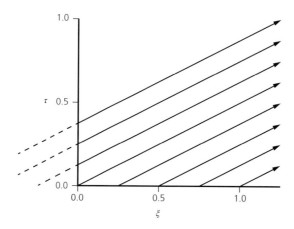

Figure 8.5 Plot of characteristics for the advection equation, with $\Lambda = 2$ (as an example). The value of θ_N at ξ_0 is preserved along the entire length of the characteristic, but the ξ at which that characteristic occurs translates with time τ, as shown schematically for some sample characteristics. Some characteristics are drawn with a dashed extension to the left of the τ-axis to illustrate that characteristics that come into positive values of ξ at times $\tau > 0$ originate on the negative half of the ξ-axis. The characteristics are all straight lines in this example, because the velocity (in this case $\Lambda = \lambda/\alpha$) is constant.

Using the symbol $\mathcal{H}$ for the Heaviside step function,[6] we can write the initial condition for θ_N as

$$\theta_N(\xi, \tau = 0) = \mathcal{H}(-\xi), \tag{8.89}$$

or the solution for all times as

$$\theta_N(\xi, \tau) = \mathcal{H}(-\xi_0 - \Lambda\tau). \tag{8.90}$$

Figure 8.6 shows a schematic/perspective rendering of the solution in θ_N, ξ, τ-space. The nearest, dark-gray panel, located at $\xi \leq 0$, $\tau = 0$, represents the initial condition; θ_N is equal to 1 for negative ξ, and 0 for positive ξ values. As time increases (progressively lighter gray panels), the location of the front moves further to the right (toward more positive values of ξ). The characteristics (arrows drawn in the ξ, τ plane) track the locations of each initial ξ_0 point from the ξ-axis at $\tau = 0$ as they translate in time and space.

Although our solution to the advection equation neglects the early time transients in fluid velocity, we can use our approximate solution as an aid to understanding conceptually what the solution to the full, variable velocity advection equation would be. For one thing, we know that the constant velocity equation gives straight characteristics, because points initially at some ξ_0 translate to larger values of ξ at a constant rate as time increases (i.e., at a constant

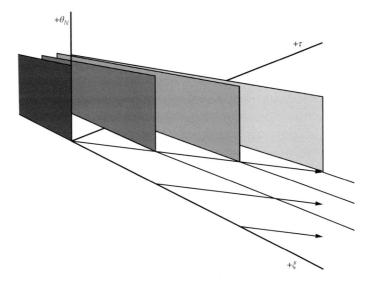

Figure 8.6 Perspective diagram showing the method of characteristics solution to the advection equation problem (Eq. 8.65). The solution, given by Equation 8.90, is shown in θ_N, ξ, τ-space, where θ_N is the dimensionless concentration of transported quantity (in this problem, nitrate), ξ is the normalized distance, and τ is the nondimensional time. The gray panels, which mark the location of the advecting front, can be seen to track toward progressively greater values of ξ as time increases.

velocity); the slope of the characteristics is governed by the Peclet number (in this case, the modified Peclet number, Λ), which functions as a dimensionless velocity. Since the velocity determines the slope of the characteristics, a temporally changing velocity will impart a curvature to the characteristics (i.e., a changing slope). However, in our problem the front will always retain the shape of a unit step function, with a zero concentration of nitrate ahead, and the maximum concentration behind the front. Some problems represented by the advection equation allow for the formation of "shock waves," where characteristics cross or intersect; for example, in compressible flow when the speed of sound is exceeded (Mach numbers >1) or for open channel flows transitioning between subcritical and supercritical flow (Froude numbers <1 or >1, respectively). In these cases, the solutions to the advection equation become multi-valued, and solutions in the traditional sense no longer exist. We will not have this problem here, but the sharp front will cause other difficulties when using finite differences to find numerical approximations of the solution.

8.7.3 Qualitative behavior of the ADE

With what we have learned about the behavior of the various special cases of the ADE, we can now discuss qualitatively how we would expect the solution of our problem to look. It remains to be seen whether or not we will be able to answer quantitatively the original question (i.e., what is the concentration of nitrate in King Augeus's well?), but at least we can identify the most important factors that govern the answer, and give a convincing description of the way in which the system is likely to work.

Probably, the first question we need to answer is, what is the magnitude of the Peclet number? As we have seen in the preceding sections, the value of the Peclet number is critical for determining what behavior the system will demonstrate. From Equation 8.42, we know that the Peclet number for our problem is defined as

$$\lambda = \frac{U_c L}{D_{AB}},\tag{8.91}$$

while, from Equation 8.36, the characteristic velocity U_c is given by

$$U_c = \frac{K(H_0 - H_L)}{\phi L}.\tag{8.92}$$

Substituting, we find that

$$\lambda = \frac{U_c L}{D_{AB}} = \frac{K(H_0 - H_L)}{\phi L}\frac{L}{D_{AB}},\tag{8.93}$$

$$\lambda = \frac{K(H_0 - H_L)}{D_{AB}\phi}.\tag{8.94}$$

The diffusivity of dilute nitrate in water, from laboratory experiments, is about $2 \times 10^{-9}\,\mathrm{m^2/s}$ ($1.92 \times 10^{-9}\,\mathrm{m^2/s}$, according to Hill (1984) and Parson (1959)). The hydraulic conductivity of sandstones is more variable, running roughly in the range of 10^{-6}–10^{-10} m/s (Freeze and Cherry, 1979) for unfractured rock. For now, we can pick a value from the middle of the range as $K = 10^{-8}$ m/s as a reasonable hydraulic conductivity. Similarly, ϕ is variable for different sandstones—although not as variable as the hydraulic conductivity—and can run between perhaps 0.05 and 0.3. We can let $\phi = 0.2$ as a preliminary guess, and we can feel confident in this quantity, because the small amount of variability won't make a huge difference (a factor, rather than an order of magnitude) on the Peclet number we calculate.

The remaining parameter we need in order to estimate the Peclet number is the head differential that drives flow from the boundary at $\xi = 0$ to $\xi = 1$. We don't know the magnitude of $H_L - H_0$ in this case, but we will assume the change in head due to Hercules's activities was 10 m. We can therefore estimate the Peclet number to be on the order of

$$\lambda \approx \frac{10^{-8}\,(10)}{(2 \times 10^{-9})\,0.2}, \tag{8.95}$$

$$\lambda \approx 250. \tag{8.96}$$

On the basis of this calculation, we can reasonably neglect any contribution from the diffusion term of the ADE, which is (roughly) 1/250th the magnitude of the advection term. By this measure, the advection equation (Eq. 8.61) would be a good representation of the nitrate transport problem at the Augean Stables, and we could probably count on concentrations of nitrate in King Augeus's well that are on the same order as the concentrations in the infiltrated water (i.e., with little contribution from diffusion, the contaminant front will be sharp, with "clean" water ahead of the front, and the water behind the front having a nitrate concentration of approximately $\theta_N = 1$). If we take the thickness of the aquifer to be $b = 20$ m, and the storativity to be 10^{-5}, we can also estimate α as

$$\alpha \approx \frac{D}{D_{AB}} = \frac{T}{S D_{AB}} = \frac{K b}{S D_{AB}}, \tag{8.97}$$

$$\alpha \approx \frac{10^{-8}\,(20)}{(10^{-5})\,2 \times 10^{-9}} = 10^7. \tag{8.98}$$

Since we know the well is located at $\xi = 0.5$ (see Figure 8.1), if we are content with an approximate arrival time based on our solution to the constant velocity equation, we can estimate an arrival time of the nitrate concentration front at the well from Equation 8.87 as

$$\tau = \frac{\alpha}{\lambda}\,(\xi - \xi_0), \tag{8.99}$$

$$\tau = (0.5 - 0)\,(4 \times 10^4) = 20,000. \tag{8.100}$$

This may or may not be a long time, depending on the value of the characteristic time for the problem. For the numbers we have chosen for this example and assuming a length scale of $L = 100$ m, from the definition of the Fourier number, we have

$$t = t_c \tau = \frac{L^2}{D} \tau = \frac{L^2 S}{K b} \tau = \frac{10^4 \, 10^{-5} (2 \times 10^4)}{10^{-8} \, 20}, \tag{8.101}$$

$$t = 1 \times 10^{10} \, \text{s}, \tag{8.102}$$

or about 317 years. From this calculation, it looks as though King Augeus has little to worry about; however, we have used the lowest possible velocity (remember that our solution to the advection equation was based on the steady-state, characteristic velocity), and the actual velocities—at least at early times—will be a bit higher, with a correspondingly lower travel time. Of course, we could solve the advection equation with variable velocity,

$$\frac{\partial}{\partial \xi} [v \theta_N] + \frac{\alpha}{\lambda} \frac{\partial \theta_N}{\partial \tau} = 0, \tag{8.103}$$

in order to get a more realistic estimate of travel time than 300 years. This could be accomplished with a numerical simulator or, potentially, by analytical methods. If we were hired as consultants to find the travel time of a contaminant front to a well, this is certainly what we would have to do.[7] However, given the scant amount of data we have about our model domain, we need to ask ourselves if obtaining a "more accurate" estimate of travel time is really a worthwhile undertaking, or if we would just be wasting our time.

Rather than worrying about the actual travel time of the nitrate to King Augeus's well, it would be interesting to review what we have learned about the behavior of a transported contaminant (i.e., nitrate) from our investigation of the limiting cases of the ADE. Looking at Figure 8.6, we can see that the advection terms in the ADE give us a sharp front of contaminant moving through the system. On the other hand, Figure 8.4, which shows the behavior of the diffusion terms, gives a more "stretched out," sloping profile for the distribution of the contaminant plume. If we combine these two very different representations of contaminant transport, we can see that the front moves through the model domain as a combination of these behaviors, with the first-order term in Equation 8.43 (the advection term) pushing the average location of the front ahead at the fluid velocity, while the second-order term (the diffusion term) diffuses the front, making it less sharp and more spread-out with time. This behavior is shown schematically in Figure 8.7.

The schematic representation of the ADE in the figure suggests another way to think about the scaling between advection and diffusion, as quantified by the Peclet number: the Peclet number gives some indication of how long it will take for the effects of diffusion to "catch up" with advection. If the Peclet

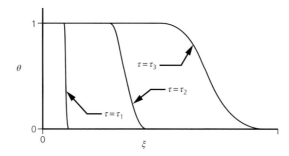

Figure 8.7 Schematic diagram showing the translation and diffusion of an advecting and diffusing contaminant front with increasing time. The curves labeled $\tau = \tau_1$, $\tau = \tau_2$, and $\tau = \tau_3$ indicate three sample times, with $\tau_1 < \tau_2 < \tau_3$.

number is large, the contaminant front will be moving rapidly and diffusion will be acting so slowly (in a relative sense) that the smearing of the front that results from diffusion won't be noticeable over the length- or timescale of our problem. The front will remain sharp, and the effects of diffusion can be neglected. Alternatively, if the Peclet number is small, the effect of smearing due to diffusion dominates, and the minor front advance that results from advection is negligible on the length- and timescales of interest. In the intermediate case (when the Peclet number is on the order of one), the center of the advancing front is described by the advection portion of the ADE but, rather than presenting a sharp aspect, the front is smeared out due to the influence of the diffusion term (because diffusion is acting on similar time- and length scales to advection). Thus, both terms are required to adequately describe the evolution of concentration in the aquifer with time and distance. This kind of "composite" behavior can be visualized by conceptually adding the diffusion solution, shown graphically in Figure 8.4, and the (simplified) advection solution, shown in Figure 8.6.

8.8 Cauchy conditions

If you've been thinking deeply about the qualitative behavior of the ADE, you may have noted a problem with the way we have specified our boundary conditions. Recall from Equation 8.46 that we have specified the dimensionless concentration of nitrate to be zero at the boundary $x = L$; that is, $\theta_N(\xi = 1, \tau) = 0$. This works fine as long as the advancing concentration front is far from the boundary; however, look carefully at the possible solution space in Figure 8.3. In this figure, all dimensionless concentrations $0 \leq \theta_N \leq 1$ are allowed at the boundary at $\xi = 1$. This can't happen if we set the boundary condition to $\theta_N = 0$. So, how can we have nitrate concentrations at the boundary that are greater than zero, while maintaining a zero concentration at the boundary?

This wasn't a problem with the advection equation, because we had to drop one of the boundary conditions to keep the first-order PDE from being overspecified. If we are going to solve the second-order ADE, however, we must have two boundary conditions (as well as an initial condition), but we can no longer specify the boundary at $\xi = 1$ to have a zero concentration, or we will not get physically realistic solutions. Furthermore, we might be interested in estimating, for example, the rate at which nitrates are discharging from the aquifer to the River Peneus, but if we set the concentration at the boundary equal to zero, we won't get a realistic estimate of flux.

Although there are several ways to handle this situation, probably the most mathematically correct (as well as the most computationally expedient) is to apply our boundary condition of the third kind (i.e., a Robin, or conduction/convection boundary; see Section 5.3.3.3) at $\xi = 1$,

$$\frac{\partial \theta_N}{\partial \xi} (\xi = 1, \tau) + \beta \theta_N (\xi = 1, \tau) = 0, \tag{8.104}$$

where β is the Biot number for species transport, which quantifies the balance between the rate at which nitrate can be transported to the boundary through the aquifer and the rate at which it can be carried away by the River Peneus on the other side of the boundary. This works well as long as diffusion dominates, but when advection is significant the boundary condition of the third kind becomes nonhomogeneous, which complicates matters somewhat.

In numerical models, another common remedy for this situation is to place the boundary so far from the domain of interest that the imposed condition never influences the solution within the model domain. This works, but it is computationally costly; furthermore, if the boundary turns out to be too close to the domain (i.e., the leading edge of the concentration front reaches the boundary, even at very small concentrations), the influence of the boundary "feeds back" to the solution within the domain, and may make the calculations unreliable. The problem of how to specify a "free exit" boundary for numerical simulations of contaminant transport problems is discussed in some detail by Frind (1988). However, be aware that there is widespread confusion in terminology within the hydrologic community regarding boundary conditions of the third kind and another type of boundary condition known as *Cauchy conditions*.

In Cauchy conditions, both the flux and the potential are specified on the same boundary, *independently*,

$$\theta(\xi = 0, \tau) = \theta_0, \tag{8.105}$$

$$\frac{\partial \theta}{\partial \xi}(\xi = 0, \tau) = v_0. \tag{8.106}$$

Equation 8.105 specifies a potential of θ_0 on the boundary at $\xi = 0$. At the same time, Equation 8.106 specifies a nondimensional flux, v_0, entering the model domain at the same boundary. This independent specification of two conditions on one boundary is handy for hyperbolic equations on open domains (e.g., the wave equation), but can't be effectively applied to our contaminant transport problem because we have a finite domain.

It is very common for people to confuse Cauchy conditions with Robin conditions (the type 3 boundary condition). There is a superficial similarity between these two conditions; both sets of conditions involve a first derivative and a potential, but in Cauchy conditions, *both* the first derivative *and* the potential are specified *independently*. As a result, Cauchy conditions are *two* conditions on *one* boundary. Robin boundary conditions, on the other hand, state that the flux and potential must match across the boundary, with the balance between them weighted by the Biot number. The first derivative and potential are *not* specified independently in boundary conditions of the third kind; therefore, when a type 3 boundary condition is indicated it only serves as *one* condition, and a second condition will be required on another boundary in order to have a well-posed problem (for a second-order governing equation). These two ways of specifying boundary conditions are different, but in the groundwater community in particular, it is commonplace to use the term "Cauchy conditions," when what is really meant is "Robin" or "type 3" conditions. A quick literature search will turn up dozens of articles in highly rated journals that confuse type 3 boundary conditions with Cauchy conditions. Be sure you understand the difference, and do not contribute to the widespread confusion in terminology surrounding these conditions!

8.9 Retardation and dispersion

In Section 8.7.3, we found that a solution to the advection equation with a constant velocity predicted very long nitrate travel times from the River Alpheus to the well of King Augeus. There are several reasons why the model predicts unrealistically long travel times. Most obviously, the constant fluid velocity used in the simplified problem neglects early time transient fluid velocities, but there are other reasons as well. For example, we used an "average" value for the hydraulic conductivity of sandstone taken from Freeze and Cherry (1979). In real life, however, the value of 10^{-8} m/s used to calculate the Peclet number is probably too low, because fluid flow through real aquifers is almost always controlled by heterogeneity—joints and fractures—that are likely to give effective hydraulic conductivities two to four orders of magnitude greater than those for the bulk rock.

There are other problems with our parameters, as well. Chief among these is the value we chose for the diffusion coefficient (diffusivity). The value we used, 2×10^{-9} m^2/s (Hill, 1984), was the *binary diffusion coefficient* of nitrate in water. The binary diffusion coefficient, usually written as D_{AB} [L^2/T], describes the rate of substance A (in this case nitrate) diffusing through substance B (in this case water). In fact, the binary diffusion coefficient is inadequate to describe the diffusion of species through a porous medium. In general, the binary diffusion coefficient must be corrected for the adsorption of the transported species onto the surface of the porous medium; this correction is usually applied by means of a *retardation coefficient, R,* defined as

$$R = 1 + \frac{k_d \rho_s}{\phi},$$ (8.107)

where k_d is the "partitioning coefficient" [L^3/M] that quantifies the volume of solids required to furnish sufficient surface area and sorption sites to hold a unit mass of species at a given temperature, ρ_s is the density of the solid phase, and ϕ is the porosity. Retardation is always greater than or equal to 1 ($R = 1$ is no retardation).[8] Working together with retardation is the tortuosity, r, of the pore spaces through which the transported species must travel. Tortuosity is unitless, and takes on values between $0 < r \leq 1$, with typical values being in the range of 0.2–0.9 (Clark, 1996). The relationship between the *effective diffusivity*, D_{eff}, and the binary diffusion coefficient, retardation, and tortuosity is given by (Clark, 1996):

$$D_{\text{eff}} = \frac{D_{AB}\, r}{R}.$$ (8.108)

Because tortuosity is always $r \leq 1$ and retardation is always $R \geq 1$, these two factors decrease the effective diffusivity in comparison to the binary diffusion coefficient. This explains laboratory measurements of the effective diffusion coefficient in a porous medium, which invariably show a smaller effective diffusivity than the binary diffusion coefficient would suggest. For example, Hill (1984) found the effective diffusivity of nitrate in chalk to be on the order of $0.5 - 3.2 \times 10^{-10}$ m^2/s, as opposed to $\sim 2 \times 10^{-9}$ m^2/s for binary diffusion of nitrate in water.

Effective diffusivity, however, is only part of the story. While retardation and tortuosity are busy reducing the magnitude of the diffusion coefficient, the effective diffusivity is being augmented by a quantity known as the *mechanical dispersivity*. Mechanical dispersion is a process that causes an increasing spread of solute about the center of mass of a species pulse. Because a "slug" of solute moves along many different pathways, each of which has a slightly different travel distance from one point to the next, aliquots of fluid/species mixture tend to separate over time/distance traveled. Most transport model formulations bundle the effective diffusivity and mechanical dispersivity into a single quantity,

$$D_h = D_{\text{eff}} + D_m = \frac{r D_{AB}}{R} + D_m,$$ (8.109)

where D_m is the mechanical dispersivity and D_h is the *hydrodynamic dispersivity.*

Mechanical dispersivity and, by extension, hydrodynamic dispersivity are extremely complex quantities, both mathematically and in terms of trying to make realistic estimates for model parameterization. Dispersivity is a tensor quantity, and is a function of many factors, including the species and transporting fluid, the heterogeneity, anisotropy, and surface charge of the porous medium, and the velocity of the fluid. Longitudinal dispersivity (dispersivity in the direction parallel to flow) is usually 3–10 times greater than transverse dispersivity (dispersivity perpendicular to flow), and measurements of effective hydrodynamic dispersivity appear to be a function of the scale of measurement. Numerous approaches have been used to model hydrodynamic dispersion (for a review, see Bear (1988)). Most commonly, dispersivity is estimated on the basis of literature values, and inverse modeling is used to refine estimates from the literature.

8.10 Numerical solution of the ADE

The material in Section 8.7 has likely convinced you that the ADE is a difficult equation to solve analytically in all but the simplest of cases. Unfortunately, numerical solutions of the ADE involve their own difficulties. The advection equation, in particular, is notoriously difficult to solve numerically, partly due to the potential for "jump" phenomena (i.e., the possibility that characteristics may cross), and partially because of the presence of sharp fronts (it is very difficult to simulate numerically discontinuous functions such as the Heaviside step function). Fortunately, there is little concern in groundwater problems for subcritical–supercritical jumps, such as those seen in high-speed compressible flow or open-channel hydraulics, and the addition of the diffusion term helps alleviate some of the difficulties seen with the advection equation due to sharp fronts. Even so, finite difference methods, in particular, are prone to errors from numerical dispersion (although finite element methods are not immune), and there are other difficulties as well. Most of these problems are troublesome to deal with when writing your own simulation code, but less so when working with commercial transport simulators. In any case, an understanding of the challenges will help to minimize problems when using numerical simulation to solve flow and transport problems.

We can apply our first- and second-order finite difference operators (Eqs. 4.43 and 4.46, respectively) to the ADE (Eq. 8.43) to arrive at the finite difference approximation,

$$\frac{\theta_{i+1}^n - 2\theta_i^n + \theta_{i-1}^n}{(\Delta\xi)^2} - \lambda\frac{v_i^n\theta_i^n - v_{i-1}^n\theta_{i-1}^n}{\Delta\xi} = \frac{\theta_i^{n+1} - \theta_i^n}{\Delta\tau}, \tag{8.110}$$

where the subscript i indicates the spatial (node) index, the superscript n gives the timestep index, and $\Delta\xi$ and $\Delta\tau$ indicate the node spacing and timestep, respectively. Note that we have dropped the θ_N notation here, to avoid confusion with the sub- and superscripts.

In writing the difference version of the ADE in Equation 8.110, we have made a number of choices about which you should think carefully; in particular, the following:

- We are assuming constant nodal spacing and a homogeneous porous medium. We have made these assumptions in all our finite difference approximations thus far, but the influence of nodal spacing on the accuracy of the solution obtained is particularly important for finite difference solutions of the ADE.
- We have used the simplest transient difference scheme; that is, we have used the explicit scheme for timestepping. More generally, we could have used a fully implicit or Crank–Nicolson scheme (see Chapter 6), but chose the simpler explicit scheme to keep the notation clear. This, of course, has important implications in terms of timestep size. On the basis of your work in previous chapters, you should be able to write your own fully implicit or Crank–Nicolson-type finite difference scheme.
- In the advection term, we have made the choice to calculate the derivative at θ_i on the basis of the difference between the values of θ_i and θ_{i-1}. We could just as easily have used the difference between θ_{i+1} and θ_i. In either case, however, we are calculating the derivative "off-center" of the node we wish to estimate. This will inevitably introduce some error into our finite difference approximation. There are ways of dealing with this more effectively in numerical schemes (see, e.g., Patankar (1980)), but they require a great deal more sophistication in coding than the other problems we have presented in this text. Note, however, that the one thing we should *not* do is to calculate the first derivative at θ_i based on the difference between θ_{i+1} and θ_{i-1}. This type of approximation to the derivative is unstable (Patankar, 1980), and should not be used.
- Equation 8.110 is presented with an *a priori* known velocity function v, which can be evaluated for any desired ξ and τ. This is not generally the case; more commonly, the velocity field must be evaluated by solving, iteratively or simultaneously, the transient diffusion equation. The need to solve multiple, coupled equations considerably increases the difficulty level of writing transport code.

As I have said several times now, it is well-known that the ADE is difficult to solve numerically, and the larger the magnitude of the Peclet number (i.e., the more advection dominates the problem), the greater the numerical difficulties. Apart from all the other problems we have discussed in this chapter, the real roadblocks to numerical solution are numerical instability and numerical dispersion. Most of the other problems to which I have alluded, such as the need to solve simultaneously the flow equation and the transport equation, are more accurately described as complexity of undertaking, rather than true difficulty of solution. Complexity of undertaking is largely avoided by using ready-made numerical simulators; in contrast, you must constantly be on your guard against numerical dispersion and instability when finding numerical solutions of the ADE.

Numerical instability is a tendency to which numerical approximations to PDEs (and ODEs) are prone under certain conditions, in which successive iterations oscillate about the true solution. In mild cases, a small amount of oscillation may be acceptable, but in severe cases the numerical approximation will overestimate a value on one iteration, then compensate by underestimating on the next iteration, and so on, with the swings becoming wilder at each iteration. We have already mentioned, for example, that transient simulations using the explicit scheme must meet the requirements (Wang and Anderson, 1982):

$$\frac{T\Delta t}{S(\Delta x)^2} \leq 0.5 \; ; \; \text{for 1D,} \tag{8.111}$$

$$\frac{T\Delta t}{S(\Delta x)^2} \leq 0.25 \; ; \; \text{for 2D, } \Delta x = \Delta y, \tag{8.112}$$

where T is transmissivity, S is storativity, and Δt and Δx are the timestep and spatial discretization (i.e., the nodal spacing), respectively.

I have also mentioned several times in this chapter that finite element approximations are more robust for solving the ADE than ordinary finite difference schemes. Although this is true, finite element simulators have their own difficulties with the ADE. Perhaps most importantly, finite element models are prone to numerical oscillations and instability near concentration fronts, and these oscillations become more severe as the fronts become sharper. Although the technical reasons behind these oscillations are beyond the scope of our discussion,[9] these instabilities can be reduced or eliminated by adhering to the following restrictions (Huyakorn and Pinder, 1983):

$$\text{Pe} = \frac{U\Delta x}{D} \leq 2, \tag{8.113}$$

$$\text{Cr} = \frac{U\Delta t}{\Delta x} \leq 1, \tag{8.114}$$

where U is the velocity and D is the diffusivity ($D = T/S$).

In Equation 8.113, Pe is the *local Peclet number*; that is, a Peclet number specifically applied to individual elements in the model grid, and Cr in Equation 8.114 is the *local Courant number*, which (in essence) tracks the speed at which a front transits the gridblocks or elements. When evaluating a grid, the most restrictive (the largest) value of Δx should be chosen to calculate the local Peclet number, because these are the gridblocks that will first develop numerical instabilities. Similarly, the most restrictive (the smallest) value of Δx for a grid should be used for calculating the maximum allowable timestep size with Equation 8.114. When using either criteria on a 2D or 3D grid, the most restrictive of Δx, Δy, or Δz should be used.

Although the restrictions on grid discretization, timestep size, fluid velocity, and diffusivity given in Equations 8.113 and 8.114 are nominally intended for finite element models, they provide good guidelines for finite difference schemes, too, because they help to minimize numerical dispersion, which is the other main problem with numerical approximations of the ADE. Numerical dispersion, as opposed to the hydrodynamic dispersion discussed in Section 8.9, can be defined as follows:

Numerical dispersion: artifact dispersion-like behavior that arises as a result of errors, inaccuracies, or insufficiencies in the process of a numerical approximation.

When using a finite difference scheme to find an approximate solution to a governing equation, numerical dispersion can arise from overly coarse grid discretization, too large a timestep size, "loose" convergence criterion, or finite-precision math errors (i.e., round-off error). The effect of numerical dispersion is to introduce greater than physical spreading of the dependent variable (species, energy, or, in the case of the transient diffusion equation, the head/pressure distribution). Sharp concentration fronts are especially prone to numerical dispersion; a front that should be a step function will generally end up dispersing to a more gradually sloping profile unless very fine grid discretization and small timesteps are enforced. Even moderate amounts of numerical dispersion lead to mass balance errors (the apparent loss of mass from the simulation due to accumulation of calculational errors), and in severe cases the transported quantity will tend to vanish mysteriously from the model domain. Although numerical dispersion can happen in any transient equation (e.g., the transient diffusion equation, the ADE, or the advection equation), it is particularly common in the transport equations (the advection equation and the ADE). As a result, it is important to always do a mass (and/or energy) balance calculation on your numerical results. If the mass present in the model domain at the end of the simulation is within 5% of the initial mass plus all mass entering the domain through the boundaries or source terms (including mass leaving, which is, of course, negative mass entering), the results of the calculation may be reliable. If a mass balance calculation is in error by more than 5%, numerical dispersion is a likely culprit, and the simulation results cannot be trusted. In this case, try refining the grid, reducing the maximum timestep size, or using higher precision arithmetic in your simulation code.[10]

Although finite element schemes are less prone to numerical dispersion than finite difference schemes, the numerical instability (undershoot/overshoot) that is associated with too-coarse gridding ($Pe \geq 2$) or timestep sizes that result in a Courant number $Cr \geq 1$ can create a situation where a modeler must strike a balance between numerical instability and practical modeling necessity. Particularly, in transport simulations with sharp concentration fronts and/or high fluid velocities, the discretization and timestep size required to avoid numerical instability may become too restrictive, either requiring too many elements to be computationally tractable, too small a timestep to allow the calculation to complete in a reasonable time, or both. In these cases, it is common to violate, for example, the local Peclet number criterion by a factor of 5 or even 10, and switch to upstream weighting in preference to harmonic weighting (see Chapter 9) between elements. Upstream weighting suppresses numerical instability, but introduces numerical dispersion (Huyakorn and Pinder, 1983). Playing these types of games with your code will help you to get the most performance from your simulation, but be aware of the consequences of making trade-offs, and be certain to always complete a mass balance calculation on your simulation results.

8.11 Conclusions

In this chapter, we have examined the transport equation, more commonly known as the advection-dispersion equation (ADE), convection–diffusion equation (CDE), or the Fokker-Planck equation. The ADE is a much more complex and difficult equation to solve than the others we have worked with, but is indispensable for modeling the transport of mass (species) and energy in porous media. In addition to a transient (temporal) derivative (often called the "accumulation term"), the ADE has a first derivative that represents transport by advection and a second derivative that represents diffusive (dispersive) processes. While nondimensionalizing the ADE, we encountered an old friend, the Fourier number (dimensionless time), and found a new dimensionless parameter, the Peclet number, that quantifies the relative strength of advection over diffusion. From an understanding of the magnitude of the Peclet number and the accuracy requirements of our situation, we can decide if we are justified in neglecting either the diffusion or the advection term of the equation for our particular problem.

Using the Peclet and Fourier numbers, we examined several simplifications of the ADE, including steady-state advection and diffusion-only solutions, and a transient diffusion-only solution. Although finding a solution of the advection equation for transient fluid velocities is more analytically demanding than we can address in this text, we did apply the MOC to the solution of the transport equation for a contaminant front moving at a constant velocity. We also discussed a "new" (to us) set of boundary conditions, known as Cauchy conditions, in which the potential and gradient of the dependent variable are specified on the boundary, and we made special note of the fact that Cauchy conditions are different from, and should not be confused with, Robin, or type 3 conditions. Finally, we examined some of the difficulties with numerical solutions of the ADE. In particular, we found that numerical dispersion and computational instability are serious difficulties that must be overcome whenever a numerical solution of the ADE is attempted. To minimize these problems, both timestep size and grid discretization must be kept within strict limits (the local, or grid, Peclet number should be less than or equal to 2 and the Courant number should be 1 or less). These restrictions will avoid numerical instability when using finite element schemes and minimize numerical dispersion with finite difference schemes; however, they are not guarantees against numerical dispersion, and a savvy modeler will always perform a mass balance calculation before relying on the results of a simulation.

In short, working with the ADE is far more complex and difficult than the other equations we have examined in this text. We have only been able to touch on a few ideas as an introduction to the subject; in fact, entire books have been written on transport modeling, and many of the topics we mentioned in a few sentences (e.g., dispersion, retardation, and tortuosity) deserve chapters, or even books, of their own. Numerical simulation of transport processes is also a serious undertaking. Most transport problems are better handled using more general methods than the simple schemes we have been using; in particular, with either integrated finite differences (Patankar, 1980) or finite elements (e.g., Huyakorn and Pinder (1983)). If you find yourself working with transport problems, as most groundwater hydrologists eventually do, you should investigate these methods very carefully or use a commercial transport code to simulate your problem. Of course, using a commercial code to simulate a transport problem is no guarantee of a reliable result; you should use your intuition and think about the results you expect to find, both from physical and mathematical points of view, and whenever possible check your results with analytical solutions. The material presented in this chapter is the barest outline of a topic that is demanding, challenging, and rewarding. As with any undertaking, be sure to prepare well for your journey!

8.12 Problems

1. Using the MOC, find the solution to the equation:

$$\frac{\partial \theta}{\partial \xi} + \frac{1}{\xi} \frac{\partial \theta}{\partial \tau} = 0, \qquad (8.115)$$

where the initial and boundary conditions are described by

$$\theta(\xi, \tau = 0) = \mathcal{H}(-\xi) \qquad (8.116)$$

2. Given a fluid velocity of 10^{-2} m/s and a species diffusivity of 10^{-8} m^2/s, what is the largest grid spacing that can be used for a fully implicit finite difference solution to a two-dimensional problem without the potential for numerical instability or dispersion? For the grid spacing calculated, what is the largest timestep size that can be accommodated without numerical instability or numerical dispersion? Would your answers change if you were using an

explicit finite difference scheme or a finite element simulator? If so, to what values would they change?

3. Find values for the following quantities from a reliable source—a handbook of physical constants, a textbook, or a refereed journal article. Cite the source for your values.

   (a) Thermal diffusivity of water.

   (b) Coefficient of binary diffusivity for sodium chloride in water.

   (c) Thermal diffusivity of sandstone (cite a range, or give a value for a specific type of sandstone if you can find one).

   (d) Coefficient of binary diffusivity for Neptunium-237 in water.

   (e) Hydrodynamic dispersion for Neptunium-237 in saturated volcanic tuff (this one may be difficult to find; try the USDOE Yucca Mountain Project literature).

   (f) Retardation coefficients for Lithium and Bromine ions; you may take your pick of media types (retardation coefficients are specific to particular rocks or soils), or use the values from the Cape Cod tracer tests (e.g., Davis et al., 2001).

Notes

1 Because they allow for changes in the mass or energy stored in the CV, temporal derivatives are sometimes called the "accumulation term." This terminology may be heard in connection with any transient equation, not just the ADE.

2 In the general case, the molecular diffusion coefficient is a function of concentration. This is because, when a species reaches high concentrations in a fluid medium, the species diffuses out toward regions of lower concentrations, while the fluid is diffusing in the opposite direction toward regions of higher species concentration (but lower fluid concentration). For our problem, we are assuming the concentration of nitrate is too low to worry about this "back diffusion" of water; thus, we can consider the diffusion coefficient to be a constant.

3 The background concentration, C_L, normalized by the concentration range, $C_0 - C_L$, multiplies the velocity and gives rise to an additional term in the advection derivative.

4 Typically, values of $\lambda \leq 0.3$ are considered "much less than 1," but the choice of cut-off is at the discretion of the user. If there is a very small tolerance for error a criterion of $\lambda \leq 0.1$ or less may be appropriate; if some additional error is less important than ease of solution, the user may decide on a value of $\lambda \leq 0.5$, or perhaps even greater.

5 We could have saved ourselves some trouble here if we had noted earlier that the dimensionless velocity must be a constant for the steady-state problem.

6 The Heaviside step function, also called the "unit step-function," is a function $\mathcal{H}(x)$ that is equal to 1 for all values of its argument $x > 0$ and equal to 0 for all $x < 0$ (Bracewell, 1986).

7 If we were actually hired consultants, we would have to do much more besides. For example, in addition to getting better parameter estimates, we would have to take into account the gradients induced by a pumping well, and so on.

8 It is believed that, in some cases, ions sorbed onto colloids can be transported in fractured rock systems at rates greater than the average groundwater velocity. In these cases, it could be said that retardation is less than 1, but this type of situation isn't likely to occur in ordinary transport problems. It is most commonly associated with radionuclide transport; for example,

colloids were blamed for the rapid transport and subsequent early detection of plutonium in wells on the Nevada Test Site (Kersting et al., 1999).

9 According to Huyakorn and Pinder (1983), the numerical overshoot and undershoot can be explained in terms of a Fourier series expansion of the solution. The phenomenon of "ringing" (Gibbs phenomenon) is a well-known consequence of using Fourier series to represent an equation with a discontinuity (e.g., a step function).

10 All numerical simulators should use double-precision arithmetic. Round-off errors associated with single-precision arithmetic very quickly render useless all but the simplest calculations.

References

Bear, J. (1988) *Dynamics of Fluids in Porous Media*, Dover Publications, New York.

Bracewell, R.N. (1986) *The Fourier Transform and Its Applications*, revised 2nd edn., McGraw-Hill, Inc., New York.

Clark, M.M. (1996) *Transport Modeling for Environmental Engineers and Scientists*, John Wiley & Sons, Inc., New York.

Davis, J.A., Hess, K.M., Coston, J.A., Kent, D.B., Joye, J.L., Brienen, P., and Campo, K.W. (2001) Multispecies reactive tracer test in a sand and gravel aquifer, Cape Cod, Massachusetts, Part 1: Experimental design and transport of Bromide and Nickel-EDTA tracers, *Tech. Rep. EPA/600/R-01/007a*, U.S. Environmental Protection Agency, Subsurface Protection and Remediation Division, National Risk Management Research Laboratory, Ada, OK.

Freeze, R.A. and Cherry, J.A. (1979) *Groundwater*, Prentice Hall, Englewood diffs.

Frind, E.O. (1988) Solution to the advection-dispersion equation with free exit boundary. *Numerical Methods for Partial Differential Equations*, **4** (4), 301–313.

Hill, D. (1984) Diffusion coefficients of nitrate, chloride, sulphate and water in cracked and uncracked chalk. *Journal of Soil Science*, **35**, 27–33.

Huyakorn, P.S. and Pinder, G.F. (1983) *Computational Methods in Subsurface Flow*, Academic Press, New York.

Istok, J. (1989) *Groundwater Modeling by the Finite Element Method*, no 13 in Water Resources Monograph, American Geophysical Union, Washington, D.C.

Kersting, A.B., Efurd, D.W., Finnegan, D.L., Rokop, D.J., Smith, D.K., and Thompson, J.L. (1999) Migration of plutonium in ground water at the Nevada Test Site. *Nature*, **397**, 56–59, doi:10.1038/16231.

Parson, R. (1959) *Handbook of Electrochemical Constants*, Butterworths Scientific Publications, London.

Patankar, S.V. (1980) *Numerical Heat Transfer and Fluid Flow*, Routledge, Taylor & Francis Group, New York.

Polya, G. (1973) *How to Solve It: A New Aspect of Mathematical Method*, 2nd edn., Princeton University Press, Princeton.

Wang, H.F. and Anderson, M.P. (1982) *Introduction to Groundwater Modeling: Finite Difference and Finite Element Methods*, Academic Press, San Diego.

CHAPTER 9

Heterogeneity and anisotropy

Chapter summary

In previous chapters, we have made the assumption that we were working with porous media that were homogeneous and isotropic. The assumptions of isotropy and homogeneity are quite common in groundwater models, as well as in models of other physical systems, owing to the tractability of the equations that result from these assumptions. In reality, it is likely that homogeneity and isotropy do not actually exist in nature—at least not in a strict sense. The more rigorous equations of potential flow that include the effects of heterogeneity and anisotropy are much more difficult to work with than their simplified relatives, since hydraulic conductivity, permeability, diffusivity, and so on, become second-rank tensors (for anisotropic materials) and may be functions of location (for heterogeneous materials). There are many methods for "glossing over" these difficulties; that is, methods have been proposed for finding "effective" properties, for reducing the dimensionality of the permeability tensor, or for incorporating anisotropy and heterogeneity into numerical solution methods. These methods vary in their rigor, complexity, and the quantity of data needed for their application. We will look briefly at some of the simplest methods, point the way toward more complex approaches for the interested reader, and finish with some comments on when it is or is not appropriate to consider the effects of heterogeneity and anisotropy.

9.1 Understanding the problem

To this point in the text, we have generally made use of the very convenient assumptions that our model domains are characterized by materials that are *isotropic* and *homogeneous*. Most scientists and engineers are familiar with these terms; however, for the sake of clarity, we will define them here.

Homogeneity: The characteristic of being invariant to spatial translation. A property is said to be homogeneous if it is found to be unchanged when measured at different spatial locations. Materials that are not homogeneous are *heterogeneous* materials; that is, the property is found to possess different values when measured at different spatial locations.

Models and Modeling: An Introduction for Earth and Environmental Scientists, First Edition. Jerry P. Fairley.
© 2017 John Wiley & Sons, Ltd. Published 2017 by John Wiley & Sons, Ltd.
Companion website: www.wiley.com/go/Fairley/Models

Isotropy: The material property described is independent of the direction in which it is observed. An *anisotropic* material displays *differing* values of a property when measured in different directions; that is, an anisotropic material has properties that are direction-dependent.

The assumptions of homogeneity and isotropy are commonly made because they result in a very substantial simplification to the potential flow equations. Take, for example, the case of 1D, steady flow in a heterogeneous aquifer. The governing equation is

$$\frac{d}{dx}\left[K(x)\frac{dH}{dx}\right] = 0. \tag{9.1}$$

In our prior investigations of this situation (the Elysian Fields problem in Chapter 4), hydraulic conductivity, K, was not a function of position, x. As a result, we could remove K from under the derivative, dividing it off against the zero on the right-hand side of the equation[1] to obtain the 1D field equation:

$$\frac{d^2 H}{dx^2} = 0. \tag{9.2}$$

If K is a function of position, $K = K(x)$, however, we can no longer remove it from under the derivative—instead, we have to deal with Equation 9.1 as it is written, which is a more complicated endeavor. In one dimension this may not be very difficult; in two or more dimensions, in transient systems, or with complexly varying properties, solving the governing equation may become very difficult indeed.

For a second example, imagine trying to calculate the flux in a 3D, anisotropic domain. Because the flux is a vector, in a 3D medium there will be three components, q_x, q_y, and q_z. The anisotropic nature of the medium will give rise to terms in each of the components that result from gradients in the orthogonal directions; the complete statement of the three components is written:

$$q_x = -K_{xx}\frac{\partial H}{\partial x} - K_{xy}\frac{\partial H}{\partial y} - K_{xz}\frac{\partial H}{\partial z}, \tag{9.3}$$

$$q_y = -K_{yx}\frac{\partial H}{\partial x} - K_{yy}\frac{\partial H}{\partial y} - K_{yz}\frac{\partial H}{\partial z}, \tag{9.4}$$

$$q_z = -K_{zx}\frac{\partial H}{\partial x} - K_{zy}\frac{\partial H}{\partial z} - K_{zz}\frac{\partial H}{\partial z}. \tag{9.5}$$

Of course, writing out all these equations (i.e., Eqs. 9.3–9.5) every time we wanted to indicate the flux vector would be a nightmare, to say nothing of taking up a huge amount of space. We have a much more compact notation to indicate the equation for calculating flux in an anisotropic material; in general, we would write

$$\bar{q} = -\bar{\bar{K}}\nabla H, \tag{9.6}$$

where the single overbar indicates a vector and the double overbar indicates a *tensor*. Using matrices, we would write the same equation as

$$
\begin{bmatrix} q_x \\ q_y \\ q_z \end{bmatrix} = - \begin{bmatrix} K_{xx} & K_{xy} & K_{xz} \\ K_{yx} & K_{yy} & K_{yz} \\ K_{zx} & K_{zy} & K_{zz} \end{bmatrix} \begin{bmatrix} \partial H/\partial x \\ \partial H/\partial y \\ \partial H/\partial z \end{bmatrix}. \tag{9.7}
$$

Whether we are using a "more compact" notation or not, however, it is obvious that working with an equation such as Equation 9.6, which is of course the same as Equations 9.3–9.5, is a much more difficult undertaking than we have attempted so far.

From these two examples, it should be clear why groundwater hydrologists usually assume homogeneity and isotropy in their models. As we have said, however, most materials are not isotropic and homogeneous, so we need to understand how and when we can neglect or average-out heterogeneity and anisotropy, and develop tools to solve the more complex equations when necessary.

9.2 Heterogeneity and the representative elemental volume

A closer look at any material will show evidence of heterogeneity at *some* level. Even the most apparently homogeneous crystalline material contains heterogeneities (flaws) in the crystalline structure at the molecular level. The fact that we can often gloss over these heterogeneities and treat materials as though they are homogeneous is a clue for us to look more deeply at the problem. The question of when we can assume homogeneity and when we must represent some level of heterogeneity is a critical issue that modelers always need to consider when formulating their models.

To explore this concept more easily, we can use the determination of porosity as an example. Imagine we wish to determine the porosity of a (seemingly) homogeneous sandstone. Because porosity is a scalar quantity, this should be a relatively simple task.[2] We start out by taking an imaginary box, superpose the box inside of the sandstone, and calculate what fraction of the box is filled with void space. This number, in theory, is our porosity.

Suppose we start our estimate of porosity with a box (a cube) that is $10^{-18}\,\text{m}^3$ in volume (i.e., 10^{-6} m, or 1 μm on each side). In all likelihood, wherever we set our box inside the sandstone, we will get one of two possible answers: either the fraction of void space will equal 1 (we are inside a pore, so the box samples only void space) or 0 (our box is located inside a clast, and there is no void space). This is not a very satisfying answer, as we can hardly say our sandstone has a porosity of 0 (or 1). To get around this problem, we can make our box bigger; that is, we can take a larger sample size. As the sample size increases, we will get some 1s and some 0s, but at some point we will begin picking up part of a clast and part of a grain. As our box size increases, the wild swings will begin to diminish. At some point, our box will be large enough that every sample will take in several pores and several clasts, and the ratio of voids to solids will asymptotically approach a stable limit. We call this limit "the porosity"[3] of the rock. Of course, if we keep increasing the size of the box we will eventually see renewed fluctuations in our estimated value of porosity, because sometimes the box will include fractured rock, while other times no fractures are included. Similarly, cavities of various types, faults, or joints may or may not be included. Finally, if we make our box big enough, we may even begin to draw in surrounding rocks, such as limestones or shales (or other sandstones with greater or lesser degrees of cementation, fracturing, etc.), that have different porosities, and our estimates will become essentially meaningless.

If we ruminate on the results of this thought experiment, we can see that the concept of porosity isn't really valid unless we can define a scale at which the void ratio goes to a stable value. The size of the "box" for which we have a stable value of porosity (or any other parameter) is called the *representative elemental volume*, or REV. Once we identify an REV for a particular medium, we formulate our governing equations on the same scale as the REV; in other words, we treat the medium as a *continuum*, or a homogeneous material, within which we have essentially "integrated over" any discontinuities or heterogeneities. All of the models we have formulated in this text have implicitly been defined on the basis of an REV, within which we can assume the material acts as a continuum. Furthermore, it is within the continuum assumption that we must ask the following question: what qualifies as a heterogeneous medium, and what qualifies as a homogeneous medium?

9.3 Heterogeneity and effective properties

Once we understand what the REV of our model domain is, we can begin to think about whether our material has homogeneous properties or heterogeneous properties. Any heterogeneities that occur at a finer scale than the REV are, by definition, included in the continuum, so we don't have to worry about them (about which, however, more later). If heterogeneities exist within the domain on a scale that is coarser than the size of the REV, but still within the model domain, the model properties will be heterogeneous; that is, we will need to define different properties for different parts of our domain. This is another way of saying that our properties (take, e.g., hydraulic conductivity) are a function of position,

$$K = K(x, y, z), \tag{9.8}$$

where K is defined for a 3D model space.[4] Within each continuum of our model domain—that is, within each area for which we can define an REV with internally constant properties—we can define a single *effective property* that incorporates the effects of all the subscale heterogeneities into a single value (if the property is a scalar; if the property is anisotropic, the effective property will be a tensor).

9.3.1 Averaging conductivity

To see how we might be able to define a single, effective property for a continuum, we can look at some extremely simplified cases of property averaging. We will use hydraulic conductivity as an example, and we will look at two end-members that form the bounds of property averaging: flow parallel to units in a layered system, and unit-perpendicular flow in a layered system.

9.3.1.1 Flow parallel to layers

To find an effective value of hydraulic conductivity for flow parallel to layers, we can refer to a conceptual system such as the one shown in Figure 9.1. We assume the flux is in the horizontal direction, steady and uniform (i.e., parallel flow lines). For our purposes, we will assume flow proceeds from left to right (from $x = 0$ to $x = L$), although the direction of flow, as long as it is moving parallel to the layers, is immaterial to the problem. In the figure, there are four layers, each having a uniform thickness normal to the direction of flow, and each with its own hydraulic conductivity. We can assume the potential (i.e., the head) is the same for every layer at $x = 0$; similarly, the head is the same for every layer at $x = L$. We can therefore write the equation for the total discharge (per unit width) through the system as:

$$Q_{\text{tot}} = -K_{\text{eff}} \, b \, \frac{dH}{dx} = -K_1 b_1 \frac{dH}{dx} - K_2 b_2 \frac{dH}{dx} - K_3 b_3 \frac{dH}{dx} - K_4 b_4 \frac{dH}{dx}, \quad (9.9)$$

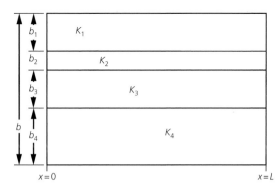

Figure 9.1 A conceptual layered porous medium for layer-parallel flow. The goal is to find a single effective conductivity, K_{eff}, that will give the same flow across the model domain (from $x = 0$ to $x = L$) as the total flow along all four layers.

where K_{eff} is the *effective conductivity*, or the conductivity that a homogeneous medium would possess if it yielded the same flux as the actual, heterogeneous system.

Since the system is assumed to be at steady state, and the change in head is the same for all layers across the domain, the factors dH/dx are all identical, and we can divide through by the gradient to obtain

$$-K_{eff}b = -K_1 b_1 - K_2 b_2 - K_3 b_3 - K_4 b_4, \tag{9.10}$$

or, solving for K_{eff},

$$K_{eff} = \frac{1}{b} \sum_{n=1}^{N} K_n b_n, \tag{9.11}$$

where N is the total number of layers (in this example, $N = 4$). Equation 9.11 is the *weighted arithmetic mean* of the hydraulic conductivities.

Equation 9.11 shows that the effective conductivity for flow parallel to layering in a "layered heterogeneous" system (i.e., a system comprising layers of different conductivity and thickness, each of which is internally homogeneous) is the thickness-weighted arithmetic average of the individual conductivities. However, this result holds *only under conditions of steady and uniform flow.* This caveat is a very important, but often forgotten, restriction on the validity of this derivation. We will come back to this point in a later section, after we have calculated effective conductivities for a few other problem geometries.

9.3.1.2 Flow across layers

For flow moving normal to bedding in a layered system, we need to use a different formulation to find the effective conductivity, K_{eff}. Referring to Figure 9.2, we can see the same four layers, with conductivities equal to K_1, K_2, K_3, and K_4, and layer thicknesses of b_1, b_2, b_3, and b_4, respectively, for a total thickness of b.

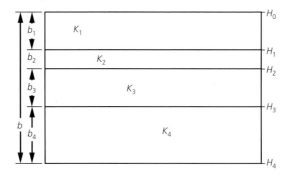

Figure 9.2 Conceptual layered porous medium for layer-perpendicular flow. We are seeking an effective conductivity, K_{eff}, that represents the conductivity of all the layers shown in this figure for flow moving perpendicular to the layers (i.e., in the z-direction).

Furthermore, as the figure shows, the heads at the upper and lower boundaries of the domain and at each boundary between layers are H_0, H_1, H_2, H_3, and H_4. The direction of flow is parallel to the z-axis.

We can write any of the fluxes across any of the layers using Darcy's law:

$$q = -K\frac{dH}{dz}. \tag{9.12}$$

In addition, since we have no sources or sink terms, and if we assume steady-state and uniform flow, we know that the gradient is a piecewise-linear function; that is, the gradient will be different across each of the layers (since they each have a different conductivity), but each individual gradient will be linear across each individual layer. As a result, we can rewrite Darcy's law for each of the layers as

$$q_1 = -K_1\frac{H_1 - H_0}{b_1}, \tag{9.13}$$

$$q_2 = -K_2\frac{H_2 - H_1}{b_2}, \tag{9.14}$$

$$q_3 = -K_3\frac{H_3 - H_2}{b_3}, \tag{9.15}$$

$$q_4 = -K_4\frac{H_4 - H_3}{b_4}. \tag{9.16}$$

Furthermore, because the system is at steady state, we know that

$$q_1 = q_2 = q_3 = q_4, \tag{9.17}$$

and that each of these fluxes is equal to the total flux across the domain. Written in terms of the desired effective conductivity, we therefore have the total flux across the domain as

$$q = -K_{\text{eff}}\frac{H_4 - H_0}{b}. \tag{9.18}$$

We can, of course, rearrange any one of Equations 9.13–9.16 to the form:

$$\frac{H_{i+1} - H_i}{q} = -\frac{b_{i+1}}{K_{i+1}}, \tag{9.19}$$

or Equation 9.18 as

$$\frac{H_4 - H_0}{q} = -\frac{b}{K_{\text{eff}}}. \tag{9.20}$$

Adding all the left sides of the rearranged equations together, and setting them equal to all the right sides of the rearranged equations gives

$$\frac{H_1 - H_0}{q} + \frac{H_2 - H_1}{q} + \frac{H_3 - H_2}{q} + \frac{H_4 - H_3}{q} = -\frac{b_1}{K_1} - \frac{b_2}{K_2} - \frac{b_3}{K_3} - \frac{b_4}{K_4}. \tag{9.21}$$

We can simplify the left-hand side of Equation 9.21 as

$$\frac{H_4 - H_0}{q} = -\frac{b_1}{K_1} - \frac{b_2}{K_2} - \frac{b_3}{K_3} - \frac{b_4}{K_4}. \tag{9.22}$$

Comparing the simplified left-hand side of Equation 9.22 with Equation 9.20, it is clear that we can write

$$-\frac{b}{K_{\text{eff}}} = -\frac{b_1}{K_1} - \frac{b_2}{K_2} - \frac{b_3}{K_3} - \frac{b_4}{K_4}. \tag{9.23}$$

Multiplying both sides by -1 and rearranging gives

$$K_{\text{eff}} = \frac{b}{\frac{b_1}{K_1} + \frac{b_2}{K_2} + \frac{b_3}{K_3} + \frac{b_4}{K_4}}, \tag{9.24}$$

or, written more compactly,

$$K_{\text{eff}} = b \left[\sum_{n=1}^{N} \frac{b_n}{K_n} \right]^{-1}, \tag{9.25}$$

where N is the number of layers (in this case, $N = 4$).

Equation 9.25 is the *weighted harmonic mean* of the hydraulic conductivities (i.e., the harmonic mean of the hydraulic conductivities, weighted by the thicknesses, b_n, of the layers). As with the case of K_{eff} for flow parallel to layers, however, we should stop to remember the restriction on this result: the weighted harmonic mean of hydraulic conductivity gives an effective conductivity for flow across layers in *a steady-state and uniform flow field*. Although this result is reasonably general, and certainly very useful (as will be seen later), these restrictions must always be kept in mind when applying the result.

9.3.2 Some statistical definitions

Before we can discuss additional methods for finding effective values of hydraulic conductivity, we first need to define a few, very basic, statistical terms. Most importantly, we have the "expected value operator," commonly denoted using angle brackets. The expected value of a random variable can be calculated for either a continuous or a discrete probability distribution. For a continuous variable X (a variable that can take on any value in a range and has a continuous pdf, $f(X)$), the expected value of X, $\langle X \rangle$, is written:

$$\langle X \rangle = \int X f(X) \, dX. \tag{9.26}$$

For a continuous variable X with a discrete probability distribution (i.e., X can take on a finite number of values in a range, with each category or "bin" having a known probability of occurrence), the expected value is given by

$$\langle X \rangle = \sum_{n=0}^{N} X_n f_n(X), \tag{9.27}$$

where X_n is the value of X for the nth bin, $f_n(X)$ is the probability of X's occurrence in the nth bin, and N is the total number of bins. The expected value of a distribution is a "measure of central tendency," or the most likely value to be picked from a random drawing.[5] We also need to define the "second moment," or *variance*, σ^2, of a distribution; for a continuous pdf,

$$\sigma_X^2 = \int (X - \langle X \rangle)^2 f(X)\, dX. \tag{9.28}$$

For a discrete distribution, the variance is given by

$$\sigma_X^2 = \sum_{n=0}^{N} (X_n - \langle X \rangle)^2 f_n(X). \tag{9.29}$$

We also need to say something about "averages." There are (at least) three ways of calculating the "average" of a dataset: the arithmetic average, X_A, the harmonic average, X_H, and the geometric average, X_G. Using the expected value operator, these three ways of calculating the average of a dataset can be written as

$$X_A = \langle X \rangle, \tag{9.30}$$

$$X_H = \left[\langle X^{-1} \rangle \right]^{-1}, \tag{9.31}$$

$$X_G = e^{\langle \ln X \rangle}. \tag{9.32}$$

We have already met two of these averaging methods (the arithmetic and harmonic means), when calculating the effective conductivity of a layered heterogeneous system for flow perpendicular and parallel to layers. If we apply the three averages to the same dataset, they will have values:

$$X_H \leq X_G \leq X_A. \tag{9.33}$$

Note that the geometric mean, X_G, is often written:

$$X_G = \left(\prod_{n=1}^{N} X_n \right)^{1/N}, \tag{9.34}$$

where Π is the usual symbol denoting multiplication of all the X_n, $1, \ldots, N$. Although this way of writing the geometric mean is, perhaps, more intuitive, you should not use this as a method for calculating the geometric mean. For any reasonable value of N, computational error (i.e., finite accuracy computations) will render the calculation inaccurate, at the very least. Although they are *mathematically* equivalent, the variation given in Equation 9.32 is a more practical method for calculating X_G.

We need these definitions for our discussion of effective conductivity, because virtually all of the conductivity results that have been obtained to date are statistical in nature. With these definitions in mind, we can go on to discuss these other situations, and the circumstances under which they may apply to various model domains.

9.3.3 Other effective conductivity results

There are a number of other, relatively simple, statements that can be made about effective conductivity, although most of the known results are "weak" (meaning they hold only in limited circumstances, have many assumptions, or give broad ranges). For example, it is known that, for uniform flow, the effective conductivity is always bounded between the arithmetic and harmonic means of the local conductivities, regardless of the dimensionality of the domain, as long as the correlation structure[6] of the medium is isotropic (deMarsily, 1986). Since the geometric mean is always between the arithmetic and harmonic means in magnitude (see Eq. 9.33), the geometric mean is often used as a first guess at the effective conductivity of a medium. Furthermore, for a 2D medium with a log-normal conductivity distribution (a log-normal distribution is one where the logarithms of conductivity have a Gaussian probability density function), the effective conductivity is the geometric mean of the conductivities (deMarsily, 1986). deMarsily (1986) also cites results for effective conductivities by several other investigators that hold for uniform flow in media with log-normal probability distributions; for 1D, 2D, and 3D flow, the effective conductivities, K_{eff}, are given by

$$\text{1D: } K_{eff} = K_G \left(1 - \sigma_Y^2/2\right), \tag{9.35}$$

$$\text{2D: } K_{eff} = K_G, \tag{9.36}$$

$$\text{3D: } K_{eff} = K_G \left(1 + \sigma_Y^2/6\right), \tag{9.37}$$

where σ_Y^2 refers to the variance of the natural log of conductivity, $Y = \ln K$. These are specific cases of the general result from perturbation theory (Rubin, 2003),

$$K_{eff} = K_G \left[1 + \left(\frac{1}{2} - \frac{1}{m}\right)\sigma_Y^2\right], \tag{9.38}$$

where m is the dimensionality (1, 2, or 3) of the model domain. Because the result given in Equation 9.38 is based on a perturbation expansion, it is generally assumed to be valid for "small" values of σ_Y^2. However, similarities between the perturbation solution and the Taylor series expansion of the exponential function e^x have led to speculation that the exact result is

$$K_{eff} = K_G \exp \left[\left(\frac{1}{2} - \frac{1}{m}\right)\sigma_Y^2\right]. \tag{9.39}$$

It would be an important result if it could be established that the exact solution for effective conductivity is given by Equation 9.39, since that would remove the limitation on Equation 9.38 that σ_Y^2 be small. Equation 9.39 has been shown to give good results in a number of specific situations; however, proof of its applicability in the general case is still required (for a discussion, see Rubin (2003)).

Of course, it is not always appropriate to treat heterogeneity by using a single, effective property. Usually, however, if spatially heterogeneous properties need to be represented explicitly, we will be required to fall back on numerical methods for representing the system. We will come back to this topic in a few sections, after we discuss the problem of anisotropy in more detail.

9.4 Anisotropy in porous media

Given the complexity of the equations that arise when dealing with anisotropic media (i.e., Equations 9.3–9.7), it makes sense to try to think of ways to reduce the difficulty of the problem. In part, this is simply motivated by a desire to minimize computational demands (or to simplify the problem to the point where it can be handled analytically). However, more fundamentally, it should be kept in mind that, in order to fill out the hydraulic conductivity tensor, we nominally have to supply nine values of conductivity (actually only six, because the tensor is symmetric). These data are rarely available, even in laboratory studies, and never with any certainty in a field situation. As a result, finding simplifications for anisotropic media is usually a matter of practical necessity.

The first thing that any modeler should do when faced with strongly anisotropic materials is to *rotate the model domain so as to align the coordinate axes with the principal directions of anisotropy.*[7] That is, orient the grid such that the *x*-axis is parallel to the direction of maximum (or minimum, or intermediate) conductivity, the *y*-axis is parallel to the direction of minimum (or maximum, or intermediate) conductivity, and the *z*-axis is parallel to the remaining conductivity direction. The reason for adopting this expediency is that *when the coordinate axes are aligned with the principal directions of conductivity, the off-diagonal terms of the hydraulic conductivity tensor are zero.* Effectively, this reduces the equations describing flow from Equations 9.3–9.5 to

$$q_x = -K_x \frac{\partial H}{\partial x}, \tag{9.40}$$

$$q_y = -K_y \frac{\partial H}{\partial y}, \tag{9.41}$$

$$q_z = -K_z \frac{\partial H}{\partial z}, \tag{9.42}$$

which is a significant reduction in mathematical complexity, and only requires three parameters to describe the medium, rather than the six needed for the full tensor.[8] Even greater simplification can be had if the direction of flow is aligned parallel to one of the principal directions of conductivity—in which case the problem reduces to 1D flow. Sadly, this is often not the case.[9]

9.5 Layered media

In Section 6.4.2, I made some disparaging remarks about the use of "layer cake" conceptual models in hydrology—despite the fact that they are used quite consistently in this text, as well as in groundwater hydrology in general. The fact is that, while not every system is adequately described as a layered system, significant simplifications result if a system can be conceptualized as being layered.

The argument for layered systems rests primarily with the definition of what might be called *hydrofacies*; that is, a group of geologic units that act similarly (i.e., either have similar hydraulic conductivity and storativity, or are "mixed" enough that they can be considered internally homogeneous in the REV sense) and are bounded by other packages of materials (other hydrofacies) that have contrasting properties. When such situations exist (e.g., in sedimentary formations and unlithified sediments overlying lithified materials), and particularly when the hydrofacies consist of laterally extensive tabular bodies, the common conceptual model of the layered aquifer/aquitard may apply. Strong contrast in properties between hydrofacies gives rise to the phenomenon of *flowline refraction*. Flowline refraction is the tendency of streamlines to change direction across boundaries of materials with contrasting hydraulic conductivities, analogous to the refraction of light across boundaries of materials with dissimilar optical properties (i.e., Snell's law). When streamlines cross a material boundary at an angle, they are refracted to a new orientation in the medium across the boundary that is a function of the angle of incidence and the ratio of the hydraulic conductivities of the media according to the relationship (Freeze and Cherry, 1979):

$$\frac{K_1}{K_2} = \frac{\tan \theta_1}{\tan \theta_2}, \tag{9.43}$$

where θ is the angle made by the flowline with the normal to the material boundary, and the subscripts 1 and 2 indicate the material referred to (see Figure 9.3).

The importance of streamline refraction is that, for situations in which the hydraulic conductivities (or permeabilities) of the media differ by two orders of magnitude or more, the streamlines tend toward nearly horizontal in the higher conductivity material, and nearly vertical in the surrounding lower conductivity units (Freeze and Cherry, 1979). As a result, flow in low-permeability confining units can often be neglected, and nominally 3D flow treated as 1D or 2D in the aquifer unit. Alternatively, in situations where flow from the confining

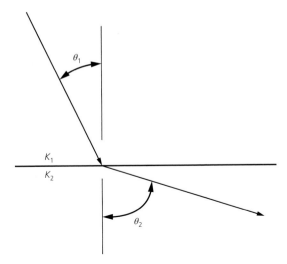

Figure 9.3 Refraction of streamlines at a material boundary. The two materials pictured, material 1 and material 2, have contrasting properties such that $K_2 > K_1$ (for the refraction pictured).

units cannot be wholly neglected, a common strategy is to treat the fluxes from overlying units as source terms rather than modeling the flow explicitly; this is the approach used in most analytical solutions for "leaky confined aquifers" (e.g., Hantush and Jacob (1955)).

9.6 Numerical simulation

Although there are many clever ways to calculate effective properties for heterogeneous materials, and many arguments for justifying the assumptions of isotropy and heterogeneity, some problems demand explicit representation of heterogeneity and anisotropy. In the cases where these realities cannot be avoided, the problems are seldom analytically tractable, and modelers will usually resort to numerical simulation. In fact, the extension of the finite difference operator to include heterogeneity and/or anisotropy is relatively straightforward. For simplicity, take the 1D, steady-state equation:

$$\frac{d}{dx}\left[K(x)\frac{dH}{dx}\right] = 0. \tag{9.44}$$

The portion of Equation 9.44 in square brackets can be written in finite difference notation as

$$K(x)\frac{dH}{dx} \approx K_{i+1,i}\left[\frac{H_{i+1} - H_i}{\Delta x}\right]. \tag{9.45}$$

Assuming constant nodal spacing (i.e., constant Δx), it therefore follows that the finite difference formulation of Equation 9.44 can be written as

$$\frac{K_{i+1,i}\frac{H_{i+1}-H_i}{\Delta x} - K_{i,i-1}\frac{H_i-H_{i-1}}{\Delta x}}{\Delta x} = 0, \tag{9.46}$$

$$K_{i+1,i}\frac{H_{i+1} - H_i}{(\Delta x)^2} - K_{i,i-1}\frac{H_i - H_{i-1}}{(\Delta x)^2} = 0, \tag{9.47}$$

$$\frac{K_{i+1,i}(H_{i+1} - H_i) - K_{i,i-1}(H_i - H_{i-1})}{(\Delta x)^2} = 0. \tag{9.48}$$

In Equations 9.45–9.48, we have used the double subscript on the conductivity K to indicate that the hydraulic conductivity is an effective property between the two referenced nodes. Solving Equation 9.48 for H_i,

$$\frac{K_{i+1,i}H_{i+1} + K_{i,i-1}H_{i-1}}{K_{i+1,i} + K_{i,i-1}} = H_i. \tag{9.49}$$

Assuming that $\Delta x = \Delta y$, the obvious extension to two dimensions is

$$\frac{K_{(i+1,i),j}H_{i+1,j} + K_{(i,i-1),j}H_{i-1,j} + K_{i,(j+1,j)}H_{i,j+1} + K_{i,(j,j-1)}H_{i,j-1}}{K_{(i+1,i),j} + K_{(i,i-1),j} + K_{i,(j+1,j)} + K_{i,(j,j-1)}} = H_{i,j}, \tag{9.50}$$

where the subscripts in parentheses indicate the nodes over which conductivity is averaged. The extension of Equations 9.49 and 9.50 to the unsteady (time-varying) case, although not difficult, is somewhat tiresome and notationally tedious; thus, I cheerfully leave it as an exercise for the student.

Despite the wearisome nature of the finite difference formulation, the application of these operators to problems of anisotropic materials is, for the most part, straightforward. For example, providing the principal directions of conductivity are aligned with the coordinate axes, it is a relatively simple matter to associate the K_x conductivity with $K_{(i+1,i),j}$ and $K_{(i,i-1),j}$, and the K_y value of conductivity with $K_{i,(j+1,j)}$ and $K_{i,(j,j+1)}$. The more difficult problem, however, is figuring out what values of K_x and K_y to use in the presence of heterogeneity. To make this clear, take the 1D finite difference operator (Eq. 9.49). Supposing we have a different conductivity assigned to each node (i.e., the gridblock surrounding node $i = 1$ has conductivity K_1, the gridblock surrounding node $i = 2$ has conductivity K_2, etc.), what values should we use for $K_{1,2}$, $K_{2,3}$, and so on?

With a little bit of thought, I hope you can convince yourself that harmonic averaging (Eq. 9.24) gives the appropriate value of effective conductivity to use between two nodes. In the case where the nodal spacing (Δx or Δy) is the same for both nodes (i.e., it is the same distance from the nodal center of each block to their common interface), Equation 9.24 reduces to

$$K_{i+1,i} = \frac{2K_{i+1}K_i}{K_{i+1} + K_i}. \tag{9.51}$$

If the nodal spacing is not constant, the full form of Equation 9.24 must be used, with the b_n values representing the distance from the nodal centers of the gridblocks to their common interface, respectively. Of course, if the nodal spacing is not constant, the formulations of the finite difference operator are not themselves valid, and the student will need to derive the appropriate finite difference operator for variable nodal spacing—however, the principle is the same.

On the basis of these considerations, we can form a strategy for incorporating the effects of heterogeneity and anisotropy into our numerical finite difference simulations. Computationally, the most efficient thing to do is to calculate an array of effective conductivity values from the nodal conductivities input by the user; these conductivities are specific to the inter-nodal "connections" (i.e., each link between two nodes has its own effective conductivity). The effective conductivities include the effects of anisotropy (by using the K_x values to calculate effective conductivities in the direction parallel to the x-axis, K_y for the y-axis, etc.), and are harmonic averages of neighboring blocks to include the effects of heterogeneity. The matrix of effective conductivities is calculated once, at the start of each simulation, to avoid the computational overhead of calculating the effective properties at each timestep. Once the conductivity matrix is filled, the simulation can proceed using the appropriate finite difference operator— either by iteration or direct solution, using an explicit or implicit formulation, and so on, just as with the homogeneous and isotropic case (although with a more complicated operator). Most of the difficulty in this situation is in bookkeeping— although this can be significant.

9.7 Some additional considerations

When developing a numerical model to represent a heterogeneous domain, it is worthwhile to remember that property transitions in layered geological media are rarely abrupt. To a certain extent, the conductivity averaging process smooths the transition between two nodes somewhat although, given the tendency of harmonic averaging to be dominated by the smallest value, the smoothing that takes place in the model may be more hypothetical than actual. Besides the question of whether or not a sharp transition is a good representation of the reality in the field, discontinuous property distributions can be the cause of convergence difficulties in some cases. Although it may not be useful or necessary to specify a gradation of properties at each transition, it may be desirable to specify local refinement to the model grid in areas of rapid shifts in media properties. Such refinement is not always possible, but it is something worth keeping in mind—particularly in simulations where you suspect numerical dispersion to be a problem, or when there are convergence issues with a particular simulation containing heterogeneous property sets. Any place in the model where there are rapid changes—abrupt property transitions, areas of high gradients (e.g., near pumping or injection wells or other source/sink terms), or discontinuities (e.g., locations where a boundary condition changes from $\theta(\tau < 0) = 0$ to $\theta(\tau \geq 0) = 1$ or similar) is a likely candidate for a refined grid.

9.8 Conclusions

In this chapter, we have examined the effects of heterogeneity and anisotropy on models of fluid flow in porous media. First, we looked at methods of averaging over heterogeneities, and identified a number of approaches for determining values of effective conductivity for certain special cases. Within some limits, these methods can often be applied to our problems. While looking at when this is and is not possible, we examined the idea of an REV and the role it plays in being able to conceptualize a porous media system as a continuum.

Although we often make the simplification that our systems are homogeneous and isotropic, not every system can be reasonably represented in such a fashion; as a result, we sometimes require numerical formulations to apply to our problems. With numerical methods, both anisotropy and heterogeneity can be incorporated into our models. It is important to remember, however, that the usefulness of these more complicated representations is only as good as the data that go into them. If the modeler feels that anisotropy and/or heterogeneity is required to adequately describe the system, there is a commensurate increase in responsibility to obtain the required data. Once the data are obtained to support the simulations, and the conceptual model is sufficiently clear to the investigator, the remaining work is mostly in the realm of bookkeeping.

9.9 Problems

1. The following table gives data on hydraulic conductivity and unit thickness for a horizontally layered sedimentary sequence:

Unit	Conductivity (m/s)	Thickness (m)
Unit A	2.0×10^{-5}	120
Unit B	5.5×10^{-5}	165
Unit C	2.7×10^{-4}	75
Unit D	1.1×10^{-7}	3
Unit E	4.4×10^{-5}	55
Unit F	8.7×10^{-4}	45

(a) What is the horizontal effective hydraulic conductivity of the units? Does any one particular unit dominate the hydraulic character of the group?

    (b) What is the vertical effective hydraulic conductivity of the units? Does any one particular unit dominate the hydraulic character of the group?

2. Using finite differences (i.e., Eq. 9.49), solve the steady-state problem of vertical flow through a layered stack of heterogeneous porous units with the properties given in the table for the previous problem. Calculate the effective conductivity of the stack using your finite difference solution as the basis of your estimated value of effective hydraulic conductivity. How does the value estimated from finite differences compare with the value you calculated in the second part of the previous problem?

3. Apply your knowledge of finite differences to derive an expression for the 1D transient finite difference operator. You may assume constant nodal spacing.

4. Apply your knowledge of finite differences to derive an expression for the 2D transient finite difference operator. You may assume constant nodal spacing, with $\Delta x = \Delta y = \Delta$.

5. Repeat your finite difference calculation from Problem 2 for the case of transient flow. Is the effective conductivity you estimate a constant, or does it change through time? If it changes through time, does it asymptotically converge on the steady-state value from the previous problem? How does the value of effective conductivity during the transient part of the problem compare with the steady-state value?

6. Using the steady, 2D finite differences operator in Equation 9.50, develop a model for a $100\,\text{m} \times 100\,\text{m}$ model domain with a parquet (checkerboard) medium of contrasting hydraulic conductivities, where each tile is $10\,\text{m} \times 10\,\text{m}$ and the two different media have hydraulic conductivities of $1.0 \times 10^{-5}\,\text{m/s}$ and $5.0 \times 10^{-7}\,\text{m/s}$. Calculate the arithmetic, harmonic, and geometric means of the hydraulic conductivities and estimate the effective hydraulic conductivity of the domain from your finite difference simulation. Which of the three means is closest to the finite difference estimate?

Notes

1 One of the students in my modeling class, J. Osterloh, coined the phrase "throwing it into the zero-abyss" to describe this operation.

2 Conceptually, anyway. In reality, the determination of porosity is a rather messy undertaking, primarily complicated by ambiguities in our definition of "porosity." For example, do we include open fractures, or not? Do we include pore spaces that are not connected to other pore spaces, or only interconnected pores? How big, or how small, a void space will we consider a "pore?" These and similar questions mean that we effectively rely on operational definitions of porosity (e.g., we use mercury porosimetry, the ratio of dry to wet density, etc.).

3 Actually, we call this limit the void ratio, $e = V_{\text{voids}}/V_{\text{solids}}$. The porosity is defined as $\phi = V_{\text{voids}}/V_{\text{sample}} = V_{\text{voids}}/(V_{\text{solids}} + V_{\text{voids}})$.

4 Of course, even in three dimensions, conductivity, or any other parameter, may only be a function of one or more of the coordinate axes. If it is independent of the coordinate axes, the medium is homogeneous.

5 Most people are familiar with the fact that there are several possible measures of central tendency. The three most common measures are the mean, the median (the 50th percentile), and the mode (the most frequently occurring value). For a Gaussian (normal) distribution these three are equal. The expected value given in Equations 9.26 and 9.27 is the mean, as you should be able to verify from examination of Equation 9.27.

6 By "correlation structure," I mean the way in which conductivity varies from place to place in the domain. The correlation structure of a medium describes the way in which hydraulic conductivity—or any other attribute—varies statistically in space (or time, if the data described are a time series).

7 This is unfortunately not always possible; for example, the model domain may contain two or more materials, the principal directions of anisotropy of which have different orientations.

8 It should be noted that, if the rotation of the coordinate axes to align with the principal directions of conductivity rotates the vertical axis out of line with the gravitational vector, the modeler will note increased dispersion in the numerical simulations. However, it is often the case (or assumed to be the case) that one of the principal axes of conductivity is vertical, and the other two are in the horizontal plane.

9 The approximation of 1D flow can most reasonably be made when the medium can be assumed to be isotropic. When the medium is anisotropic, any angular separation between the gradient and the principal directions of conductivity will give rise to off-diagonal terms in the hydraulic conductivity matrix, and thus greater than 1D flow.

References

deMarsily, G. (1986) *Quantitative Hydrogeology; Groundwater Hydrology for Engineers*, Academic Press, San Diego.

Freeze, R.A. and Cherry, J.A. (1979) *Groundwater*, Prentice Hall, Englewood Cliffs.

Hantush, M.S. and Jacob, C.E. (1955) Non-steady radial flow in an infinite leaky aquifer. *Eos, Transactions of the American Geophysical Union*, **36** (1), 95–100.

Rubin, Y. (2003) *Applied Stochastic Hydrogeology*, Oxford University Press, Oxford.

CHAPTER 10

Approximation, error, and sensitivity

Chapter summary

An important part of modeling is understanding how errors in models propagate into the results: will a small error in parameterization or input result in a large error in the model predictions? Evaluating the "sensitivity" of a model to small errors is relatively easy to do for analytical models, but it is something that can—and should—be evaluated for numerical simulations as well. As a lead-in to the topic of model sensitivity, we will discuss a method for approximating the solutions to problems that deviate slightly from a known solution, and apply these methods to the development of uncertainty bounds on model outputs.

10.1 Things we almost know

Everyone is familiar with the situation of looking at a difficult math problem, then suddenly seeing there is an easy answer. This is a familiar experience because the problems in most textbooks are of this sort: it usually turns out that the quantity in question just happens to be a perfect square, or an easy factoralization shows that a complicated-looking fraction can be simplified by dividing a common factor from both the numerator and denominator. For a concrete example, suppose you were asked to find the cube root of some large number,

$$y = \sqrt[3]{125}. \tag{10.1}$$

At first glance, this might seem to present a huge amount of work, but after a moment's thought we see the answer is 5. How did we know this? Well, our experience tells us that, when someone gives us a difficult-looking problem, there is often some trick to it. Knowing this, we try a few whole numbers that might be in the ballpark and—*voilà*—we very quickly find the answer.

Unfortunately, real life doesn't often hand us problems of this sort. In real life, we are more likely to encounter a problem such as

$$y = \sqrt[3]{126}, \tag{10.2}$$

Models and Modeling: An Introduction for Earth and Environmental Scientists, First Edition. Jerry P. Fairley.
© 2017 John Wiley & Sons, Ltd. Published 2017 by John Wiley & Sons, Ltd.
Companion website: www.wiley.com/go/Fairley/Models

which is a more difficult problem to solve. Experience with this kind of problem tells us that we are in for a whole bunch of work, probably by trial and error (although if we're smart and have a table of logarithms handy, things will go much more quickly[1]), and it is likely we'll just give up, unless there is some really important reason to solve the problem.

If we follow on from the example in Equation 10.2, however, it makes sense to ask the following question: "is there anything I know about some similar problem—any related fact—that can help me with *this* particular problem? After a little thought, we are likely to realize the answer must be a number *around* 5, since 125 is a perfect cube. The answer to Equation 10.2, therefore, must be 5 and a little. At least now we have somewhere to start.

10.2 Approximation using derivatives

So, we have somewhere to start. In mathematical terms, we say we know the solution to the zeroth order—or, more commonly, we say we know *the leading order* of the solution's behavior. In other words, we have the bulk of the answer...now we need to refine our estimate (i.e., we want to find the *first-order correction*).

How can we do this? It might not occur to you to think in terms of derivatives in this instance, but it turns out it will be helpful to us. Recall that the definition of a derivative is

$$\frac{dy}{dx} = \lim_{\Delta x \to 0} \frac{f(x + \Delta x) - f(x)}{\Delta x}. \tag{10.3}$$

In a more compact notation, we can write Equation 10.3 as

$$\frac{dy}{dx} = \lim_{\Delta x \to 0} \frac{\Delta y}{\Delta x}. \tag{10.4}$$

In fact, if we are willing to settle for an *approximation* of the derivative, we can skip the limit and write

$$\frac{dy}{dx} \approx \frac{\Delta y}{\Delta x}, \tag{10.5}$$

which holds as long as Δx is sufficiently small (this is what we do for finite differences).

Now we can use these definitions to link up derivatives and approximations. The "generic" verison of Equations 10.1 and 10.2 is

$$y = \sqrt[3]{x} = x^{\frac{1}{3}}. \tag{10.6}$$

From this, we can find the derivative

$$\frac{dy}{dx} = \frac{1}{3} \frac{1}{x^{2/3}}. \tag{10.7}$$

We know that the solution to Equation 10.2 will be *close* to the solution of Equation 10.1; that is, the solution to Equation 10.2 should be the solution to Equation 10.1 *plus a little bit*,

$$y(x) = y_0 + \Delta y = y(x = x_0) + \Delta y. \tag{10.8}$$

We already know what y_0 is (i.e., 5). We can rearrange Equation 10.5 to get a formula for Δy,

$$\Delta y = \Delta x \frac{dy}{dx}, \tag{10.9}$$

and since we know the derivative of y (from Eq. 10.7), we can put the whole thing together as

$$y(x = 126) \approx y(x_0 = 125) + \Delta x \frac{1}{3} \frac{1}{x^{2/3}} (x_0 = 125). \tag{10.10}$$

In Equation 10.10, we have evaluated the derivative at the known value of x (i.e., the value of x for which we know the value of y). To carry out the operations in Equation 10.10, we recognize that the cube root of 125 is 5, 5 squared is 25, and 25 times three is 75. Furthermore, we know that Δx, which is the difference between the known value of y and the value we want to find, is 1, so we have

$$y(x = 126) = 5 + (1) \frac{1}{75} \approx 5.0133, \tag{10.11}$$

which is identical to the actual value for the cube root of 126 to four decimal places.

10.2.1 Another example

This method of estimation works because in many situations it is easier to calculate the derivative of a function than to calculate the actual function itself. For example, what if we wanted to know the logarithm (base 10) of 10.05? The log of 10 is 1, and the derivative of $\log_a x$ is:

$$\frac{d \log_a x}{dx} = \frac{1}{x \ln a}, \tag{10.12}$$

where a is the logarithm base and $\ln a$ is the natural logarithm of a. From this, we immediately find that:

$$\log(x = 10.05) \approx \log(10) + \frac{1}{20} \frac{1}{10} \frac{1}{2.3026} = 1.0022, \tag{10.13}$$

which is correct to four decimal places. In addition, it was much simpler to obtain this estimate than to calculate the true value of the logarithm!

10.3 Improving our estimates

Thinking hard about these examples, we can see one of the weaknesses of the estimation method is that it requires Δy and Δx to be small in order to give a good approximation. The definition of "small," however, is heavily dependent on the equation being approximated. In the case of the logarithm function, a $\Delta x = 0.05$ was small enough to give an approximation for the logarithm function good to four decimal places in the vicinity of $x = 10$, but a $\Delta x = 0.5$ would yield a precision of only 3 decimal places (you should try this for yourself).

Is there anything we can do to improve our estimates when the step size is a little larger? In fact, there is. Of course, our goal in Section 10.2 was to come up with a *simple* method for getting an *approximation* of an answer. Often that's good enough for the problem at hand. However, if we are willing to put in more computational effort, we can usually improve our estimate by adding *higher order terms*; that is, additional correction terms (second order, third order, etc.). These terms follow the pattern of the first-order term; the general formula is

$$f(x) = f(x_0 + \Delta x) = f(x_0) + \frac{f'(x_0)}{1!}\Delta x + \frac{f''(x_0)}{2!}(\Delta x)^2 + \cdots, \tag{10.14}$$

where the primes indicate differentiation.

If you're really paying attention, you might recognize Equation 10.14 as the definition of the Taylor series.[2] A Taylor series expansion exists for every *analytic function*[3]; as a result, any analytic function can be represented to arbitrary precision with a sufficient number of terms in the series.[4] In the limit of an infinite number of terms, the Taylor series is the same as the function itself, and is therefore an alternative definition of the function.

Most of this is an aside, and does not directly concern us here. However, for the aspiring modeler, it is worth understanding a little bit about Taylor series because of their role in the investigation of both linear and nonlinear equations. As an example, consider the equation for the motion of a simple pendulum,

$$\frac{d^2\theta}{dt^2} = -\frac{g}{L}\sin\theta, \tag{10.15}$$

where t is time, g is the acceleration of gravity, L is the length of the pendulum, and θ is the angle of the pendulum's deviation from the vertical. Equation 10.15 is a nonlinear equation that requires the use of elliptic integrals to solve. If we know the Taylor series expansion of the sine function,

$$\sin\theta = \theta - \frac{\theta^3}{3!} + \frac{\theta^5}{5!} - \frac{\theta^7}{7!} + \cdots, \tag{10.16}$$

we can substitute it into Equation 10.15 for $\sin\theta$. However, if we can be satisfied with small θ (i.e., if we restrict the pendulum to small deviations from vertical), we can make do with only the first term of the Taylor series for sine. Then Equation 10.15 becomes

$$\frac{d^2\theta}{dt^2} \approx -\frac{g}{L}\theta, \tag{10.17}$$

which is a much easier equation to solve, but it still gives a very good approximation of the behavior of a pendulum for most applications.[5]

10.4 Bounding errors

Given the widespread availability of calculators and computers, the use of derivatives to make estimates is mostly of historical interest (unless, like me, you're fascinated with this stuff). Beyond estimation, however, the technique of approximation presented in Sections 10.2 and 10.2.1 is useful in other applications. In particular, estimation using derivatives is very helpful in analyses of the propagation of errors in a model.

To give a physical example, imagine trying to estimate the volume and surface area of a cube. To accomplish this task, you would of course begin by taking measurements of the dimensions of the cube. Because this is a real cube (as opposed to a hypothetical perfect cube), each edge of the cube would have a slightly different length; furthermore, there is invariably a certain amount of estimation error associated with each measurement taken. If we make three independent measurements on each edge of the cube, we end up with 12×3, or 36 measurements. Each of these measurements is, in principle, one realization of L, the edge length of an ideal cube. Our models of the surface area (A) and volume (V) of a cube, respectively, are

$$A = 6L^2, \tag{10.18}$$
$$V = L^3. \tag{10.19}$$

The question, then, is, *given the uncertainty of our input data, how much confidence can we place in our model output?*

10.4.1 Data uncertainty
In any investigation that involves the collection of data, the investigator commonly has to have some means of communicating to the audience (and of evaluating for her/himself) the magnitude of uncertainty that accompanies the reported data. For the situation presented in the example, evaluating the uncertainty in L is a relatively straightforward affair: the investigator will calculate the mean and standard deviation of the data, then report the mean (the *expected value*), and follow that value with a range ("plus or minus") that represents the spread of the data about the mean. The range that is reported is somewhat variable, but is commonly one or two standard deviations, or perhaps $1.96\sigma/\sqrt{N}$ (95% confidence interval of the standard error of the mean, where σ is the standard deviation and N is the number of measurements in the sample). Whatever range is used, it should be reported along with the results. This is all that is required in the simple case used here as an example; however, the extension to more complicated cases is generally obvious. For example, if we take 10 measurements of discharge in a stream every day at noon for 1 month, we might calculate the mean and standard deviation for each day's measurements, then plot the daily means with error bars equal to 2 standard deviations for that day's data.

10.4.2 Model uncertainty

Most of us have a pretty good idea of how to report data uncertainty. The question of how that uncertainty propagates through our model, on the other hand, requires a good deal more consideration. Obviously, we can't simply report the error bounds on the data as applying to the model output; models can magnify or diminish error, and we need some method for determining how our model's outputs are affected by input and parameter uncertainties. This is where our investigations into using derivatives for approximation are going to pay dividends.

Given that lead-in, it probably isn't terribly difficult to imagine how our method of approximation can be used to investigate the propagation of error through a model. If the model input is x_0, with an uncertainty of $\pm\Delta x$, the model is $y = f(x)$, with an output of y_0 when $x = x_0$ and an uncertainty of $\pm\Delta y$, we have

$$y_0 \pm \Delta y \approx f(x_0) \pm \Delta x \frac{dy}{dx}(x = x_0). \tag{10.20}$$

To make this concrete, suppose in measuring our cube we found $\langle L \rangle = 10$ cm (the expected value, or mean, of the length of the cube's edges, was 10 cm), with a standard deviation of 0.1 cm. The standard error of the mean is

$$\text{SEM} = \frac{\sigma}{\sqrt{N}} = \frac{0.1}{\sqrt{36}} = 0.0167. \tag{10.21}$$

Multiplying by 1.96 gives the 95% confidence interval that the mean of the data is 10 ± 0.03 cm. How does this propagate into our estimates of the surface area and volume of the cube? From Equation 10.18, we have

$$\frac{dA}{dL} = 12L, \tag{10.22}$$

and from Equation 10.19, we find

$$\frac{dV}{dL} = 3L^2. \tag{10.23}$$

Applying the general formula (Eq. 10.20), we have

$$A = 6L_0^2 \pm \Delta L \frac{dA}{dL}(L = L_0) = 600 \pm 3.6 \text{ cm}^2, \tag{10.24}$$

$$V = L_0^3 \pm \Delta L \frac{dV}{dL}(L = L_0) = 1000 \pm 9 \text{ cm}^3. \tag{10.25}$$

Note that the calculated volume is a little less certain than the calculated surface area, with the surface area having a relative error of 0.6%, while the volume has a relative error of 0.9%—even though the input to both the calculations has the same uncertainty.

Of course, the models used in these examples are quite simple. Although the uncertainties in Equations 10.24 and 10.25 are only a function of a single variable (L), the usual case is that the uncertainty in model output is a function of a number of uncertain variables and parameters. When multiple variables and parameters are involved, the total uncertainty of the output is the sum of the uncertainty for each input; a simple example of this is the uncertainty in the volume of a rectangular prism, where the three edges are of different lengths: $l \neq w \neq b$. The volume is given by

$$V = lwb, \tag{10.26}$$

and the total uncertainty in the volume estimate is given by the sum of the uncertainties of each of the edge lengths,

$$\Delta V = \Delta l \frac{\partial [lwb]}{\partial l} + \Delta w \frac{\partial [lwb]}{\partial w} + \Delta b \frac{\partial [lwb]}{\partial b}, \tag{10.27}$$

$$\Delta V = \Delta l(wb) + \Delta w(lb) + \Delta b(lw). \tag{10.28}$$

Although not a proof, we can see if this gives reasonable results by checking the special case where $l = w = b = L$:

$$\Delta V = \Delta L(L^2) + \Delta L(L^2) + \Delta L(L^2), \tag{10.29}$$

$$\Delta V = \Delta L\, 3L^2, \tag{10.30}$$

which is identical to the uncertainty in the volume estimate that would arise directly from Equation 10.19.

10.5 Model sensitivity

Modelers, and scientists and engineers in general, often encounter situations in which they are less interested in knowing the exact uncertainty of output that would arise from a given uncertainty of input, and are more interested in knowing *how sensitive is a model to errors* in general. Or, to put it more plainly, is a small error in model parameterization or input data going to result in a large error in the model's predictions (output)? Alternatively, will a large error in parameterization or input data be minimized by the model, leading to a relatively small error in model predictions? Questions of this type are the focus of *model sensitivity*.[6]

10.5.1 Defining sensitivity

In order to develop a definition for "model sensitivity," we can start out by defining the *relative error* of a model's input and output. Relative error can be expressed as either a fraction or a percent; because many people are used to percent error, we will use that (although the reader will appreciate the two only differ by a multiplicative factor of 100). For the sake of specificity, we will say that our model accepts some input measurement B, and yields an output A. Continuing to use the nomenclature from the previous sections of this chapter, we can call our measured data point B_0. Because of measurement (or other) error, our measured value will be somewhat different from the true value; we will say that the measured value and the actual value differ by an amount ΔB. As a result, the true value is $B_0 + \Delta B$. It should be clear to you that the *percent relative error* (i.e., the true value minus the measured value, divided by the true value and multiplied by 100) is given by

$$\% \mathrm{RE}_B = \frac{\Delta B}{B_0 + \Delta B} \times 100. \tag{10.31}$$

Applying similar reasoning to our output (dependent) variable yields

$$\% \mathrm{RE}_A = \frac{\Delta A}{A_0 + \Delta A} \times 100, \tag{10.32}$$

where A_0 is the value of A that is output by the model, and ΔA is the difference between the true value and the model's estimate.

The question we would like to answer is as follows: "if the percent relative error in our measurement of B is small, will it result in a large percent relative error in the model estimate of A?" (We say the model estimates of A are sensitive to errors in B.) Alternatively, "if we have a large percent relative error in our measurement of B, will the resulting model estimate of A have a small percent relative error?" (We say the estimates of A are robust to errors in B.) Clearly, we are most interested in the *ratio of the percent relative errors*,

$$\frac{\frac{\Delta A}{A_0 + \Delta A} \times 100}{\frac{\Delta B}{B_0 + \Delta B} \times 100}, \tag{10.33}$$

or, more simply

$$\frac{\Delta A}{\Delta B} \frac{B_0 + \Delta B}{A_0 + \Delta A}. \tag{10.34}$$

Equation 10.34 is a formula for *the sensitivity of model output (A) to errors in model input (B)*. If the value of Equation 10.34 equals 1, then a 5% (e.g.) relative error in the model input will give rise to a 5% relative error in the model output. If, on the other hand, the value of Equation 10.34 is greater than 1, the percent relative error in the output will be greater than the percent relative error in the input (e.g., if the value of Eq. 10.34 is 2, a 5% relative error in the input will result in a 10% relative error in the output). If, on the other hand, the value of Equation 10.34 is less than 1 (suppose it is 0.5), a given percent relative error of the input (say 5%) will give rise to a smaller percent relative error in the output (in this case, 2.5%). If our model gives values of Equation 10.34 less than 1, we say the model is *robust to errors*; on the other hand, if the value we calculate for Equation 10.34 is greater than 1, we say the model is *sensitive to errors*.

If you are thinking ahead, you might be able to see that there could be quite a bit of work involved in using Equation 10.34 to evaluate the sensitivity of a model. This is because, although we know B_0 and A_0, and we may be able to estimate ΔB (using, e.g., the statistics from a set of measurements, as we did in Section 10.4.1), as a rule we do not know ΔA. We will therefore have to obtain our value of sensitivity by trial and error: putting in known pairs of B_0 and A_0, then testing ranges of values for ΔB to see what values of ΔA result. This is not only tedious, but will require great attention to detail, because it is often the case that the sensitivity of a model varies over the range of the input parameter (i.e., we may calculate a particular sensitivity to ΔB for a given value of B_0, and a different sensitivity for the same ΔB at a different B_0).

Fortunately, we may be able to take a shortcut. If we suppose the *error in the input is small*, we can approximate $\Delta A/\Delta B$ in Equation 10.34 with dA/dB; if we further assume that $A_0 \gg \Delta A$ and $B_0 \gg \Delta B$, Equation 10.34 reduces to *the equation for the sensitivity of A to small errors in B* (Meyer, 2004),

$$S(A, B) = \frac{B_0}{A_0} \frac{dA}{dB}, \tag{10.35}$$

where S is the sensitivity to small errors.

10.5.2 Example 1

Looking back at our expressions for the volume and surface area of a cube, we have

$$A = 6L^2, \tag{10.36}$$

$$V = L^3. \tag{10.37}$$

The formulas for sensitivity with respect to L are

$$S(A, L) = \frac{L_0}{A_0} 12L_0, \tag{10.38}$$

$$S(V, L) = \frac{L_0}{V_0} 3L_0^2. \tag{10.39}$$

Substituting $L_0 = 10\,\text{cm}$, $A_0 = 600\,\text{cm}^2$, and $V_0 = 1000\,\text{cm}^3$, we find

$$S(A, L) = 2, \tag{10.40}$$

$$S(V, L) = 3. \tag{10.41}$$

From the sensitivities shown in Equations 10.40 and 10.41, we can see that the sensitivity of our volume estimate is indeed greater than that of the surface area estimate. Furthermore, we can understand from this our results (from Eqs. 10.24 and 10.25) that a percent relative error of input of 0.3% yields a percent relative error of 0.6% for surface area and 0.9% for volume.

10.5.3 Example 2

For a somewhat more complicated example, take the formula:

$$\theta = \beta e^{-\alpha\tau}. \tag{10.42}$$

How sensitive is our estimate of θ to small errors in α and β? The derivatives are

$$\frac{\partial\theta}{\partial\alpha} = -\beta\tau e^{-\alpha\tau}, \tag{10.43}$$

$$\frac{\partial\theta}{\partial\beta} = e^{-\alpha\tau}. \tag{10.44}$$

Using Equation 10.43, we can find the sensitivity of θ to small errors in α to be

$$S(\theta, \alpha) = -\frac{\alpha_0}{\theta_0} \beta\tau e^{-\alpha_0\tau} \tag{10.45}$$

In our equation for the sensitivity of θ to small errors in α (Eq. 10.45), we find that the sensitivity is a function of time (τ). Note, however, that the value of θ_0 is itself dependent on when we observe it; as a result, we can replace θ_0 with Equation 10.42 to arrive at

$$S(\theta, \alpha) = -\frac{\alpha_0\tau\beta e^{-\alpha_0\tau}}{\beta e^{-\alpha_0\tau}} = -\alpha_0\tau. \tag{10.46}$$

In other words, the sensitivity of our estimate of θ to small errors in α increases linearly with time. For times,

$$\tau < \frac{1}{\alpha} \rightarrow |S(\theta, \alpha)| < 1, \tag{10.47}$$

whereas $|S(\theta, \alpha)| > 1$ for $\tau > 1/\alpha$, eventually going to infinity as $\tau \rightarrow \infty$.

For the sensitivity of θ to small errors in β, we have

$$S(\theta, \beta) = \frac{\beta_0}{\theta_0} e^{-\alpha\tau}. \tag{10.48}$$

Once again, we can replace the θ_0 with Equation 10.42 to find

$$S(\theta, \beta) = \frac{\beta_0 e^{-\alpha\tau}}{\beta_0 e^{-\alpha\tau}} = 1, \tag{10.49}$$

which tells us that the sensitivity is not, in this case, a function of time. The negative value of sensitivity in Equation 10.46 indicates that the error is inversely proportional; that is, an overestimate of α of (say) 5% will result in an *underestimate* of θ (the exact amount of the underestimate being a function of τ), whereas an overestimate of β by 5% will give rise to an overestimate of θ by 5%.

10.5.4 What good is it?

Applications for our methods of estimation and error propagation are pretty obvious. On the other hand, it may not be so easy to visualize applications for sensitivity. To what uses can we put our calculations of sensitivity?

The most obvious application for sensitivity calculations is to warn us if our model is particularly sensitive to errors of input. If we see that is so, we know (hopefully ahead of our data collection) that we need to take special care in collecting accurate data; otherwise, the sensitivity of our model is going to render our predictions useless. Alternatively, of course, we can seek a different model that is less sensitive to the types of errors that are likely to arise in our data collection. Meyer (2004) gives a very interesting and entertaining example of this use of sensitivity by contrasting astronomical models by two ancient philosophers, Aristarchus and Eratosthenes. By testing whether our model is sensitive or robust to errors, we can gain some sense of how careful our data collection procedures need to be, or if perhaps we are wasting our time completely and need to consider a different approach to our problem. In this connection, we should emphasize a particular point made by Meyer (2004): that *accuracy* and *precision* are qualities of our data (to a greater or lesser extent)[7]; however, *sensitivity* and *robustness* to errors are qualities of our model. There is no such thing as "sensitive data" or "robust data," only sensitive and robust models.

For the usual case in which our model input involves measuring several parameters or types of data, the sensitivity of our model to errors tells us where to direct our efforts. For example, if we need to measure two parameters for input to our model, and the model is sensitive to one of these parameters and relatively insensitive to the other, we should of course put the majority of our effort into making careful measurements of the parameter to which the model is more sensitive, and make do with relatively casual measurements of the parameter to which the model is insensitive. This is a question of time versus money: where do we place our effort? Obviously, the greatest gains can be made by focusing our attention on the parameters to which the model shows the greatest sensitivity. This is not, however, what always happens in practice. Without a calculation of sensitivity to guide us, it is tempting to spend a great deal of effort on measuring or calculating very precisely those parameters that *we can* measure or calculate with a high degree of precision, while making do with rough estimates of those parameters and/or data that are difficult to assess. This tendency is perhaps exemplified by efforts to calculate the value of π, which has currently been carried out to on the order of 10^{13} digits—although no scientific or engineering application calls for a precision of any greater than perhaps 40 significant figures at the most.[8]

Finally, it seems appropriate to put in a word about the "sensitivity studies" that often appear in studies that involve very poorly constrained parameters. For example, more often than not, groundwater modelers have only very weak constraints on permeability (or hydraulic conductivity) within their model domain. Since the model predictions may depend quite sensitively on the distribution of permeability, something has to be done to demonstrate the way in which the model output is a function of the uncertain parameter(s). This situation is generally approached by repeating the simulations for three or four values of permeability, and plotting the resulting values of head (or whatever the dependent variable is) as a function of changing permeability. There is nothing wrong with this approach to a "sensitivity study," and often nothing more is really required. If the value of (e.g.) permeability is bounded to within two orders of magnitude, and the calculated response of head doesn't vary that much for values of permeability over the likely range, then three or four sample calculations will probably suffice to show the major trend. This is the usual case in engineering-type studies.

On the other hand, if the goal of the work is to understand some phenomenon or site—that is, if this is a *research* study—then a much more in-depth investigation may be required. It is easy to see that an analytical statement of the sensitivity of a model output to small (or large) variations in input data or parameters contains a great deal more information about the model than a few points derived from numerical simulations. In order to obtain the same level of understanding from numerical simulations alone would require a very large number of simulations, especially since model sensitivities can vary as a function of multiple variables or parameters (e.g., as a function of time). When one considers processor (computational) time, as well as the time needed to prepare input decks and postprocess and analyze the output, a comprehensive sensitivity analysis by numerical methods may be a prohibitively expensive proposition. Since, as I have tried to show in this book, every numerical model—no matter how complex—is a quantification of equations that can be written down on a piece of paper, it may be worthwhile to spend a few minutes examining the governing equations and checking the sensitivities by hand, rather than (or at least before) running a large number of simulations. Ideally, analytical and numerical methods work together in the analysis of physical systems, and a thoughtful approach to an investigation will identify areas where one method or another will be able to make a superior contribution.

10.6 Conclusions

In this chapter, we have worked our way through a number of topics that are relevant to the evaluation of model output. To start with, we demonstrated a method for extrapolating from a known solution of an equation to arrive at an estimate of an unknown solution that differs from the known one by a small amount. As was discussed in the text, this method of approximation is, perhaps, not as important in these days of computers and calculators as it was prior to the invention and widespread availability of calculational machines; however, the method can be slightly generalized and put to use in the study of error propagation. By taking this approach, we were able to show a relatively simple method for determining the impact of data or parameter uncertainty on model outputs, and showed some illustrative examples of the way in which error bounds could be established for model predictions.

In the final sections of this chapter, we looked at ways in which a model's sensitivity to errors can be evaluated. In particular, we examined the following question: "what is the effect, in terms of percent relative error, on the output of a model, given some percent relative error of the model input (or a parameter)?" This question lead us to define "model sensitivity" in terms of the relative error of output resulting from a relative error in input. For special cases (i.e., when we can assume the input errors are small), we can simplify our expression for sensitivity, and use instead an expression for "sensitivity to small errors" (Eq. 10.35). On the other hand, for situations where small errors cannot be counted on, the full expression (Eq. 10.34) for sensitivity must be applied.

Although some of the methods discussed in this chapter may seem anachronistic, I believe it is worth a student's time to learn a bit about these topics. It is fashionable to think that every problem can be solved with numerical methods, and that analytical models, methods of approximation, and the like are little more than curiosities. My personal feeling is that these attitudes are held by those who don't understand analytical methods, and can't be bothered to learn them. With a little effort, however, the world of quantitative analysis is wide open to the modeler that understands and can use both.

10.7 Problems

1. Find an approximate value for $\sqrt{17}$. What is the relative error of your approximation in comparison to the true value?
2. The value of $\sin(\pi/6) = 1/2$, and $\cos(\pi/6) = \sqrt{3}/2$.
   (a) Find an approximate value for $\sin(7\pi/36)$.
   (b) The Taylor series expansion for the sine function is

$$\sin\theta = \theta - \frac{\theta^3}{3!} + \frac{\theta^5}{5!} - \frac{\theta^7}{7!} + \cdots . \qquad (10.50)$$

Using the Taylor series approximation, find the value of $\sin(7\pi/36)$ to four significant figures. How does the approximation you found in the previous problem compare with the true value? Give the difference in terms of absolute error and percent relative error.

3. The equation for the volume of a sphere is given by

$$V = \frac{4}{3}\pi r^3.$$

Suppose you were to measure the diameter of a sphere and found it to be 100 ± 2 cm. What is the volume of the sphere, and what are the error bounds on your volume estimate?

4. Actually measuring the radius or diameter of a sphere can be quite challenging (unless you happen to have a caliper, or a hole directly through the center of your sphere). It is generally easier to measure the circumference of the sphere, and then infer the radius or diameter. Suppose you measure the circumference of a sphere to be 251 ± 1.3 cm.

   (a) Substitute the formula relating the radius (or diameter) of a circle into the formula for the volume of a sphere (given in the previous problem) to arrive at a formula for the volume of a sphere as a function of its circumference. What is the error you expect for your calculated volume in this case?

   (b) Calculate the sensitivity of your volume estimate to small errors in the measured circumference. Would you say your model estimate is sensitive to errors in the circumference?

5. Given a cylinder of height $h = 10$ cm and radius $r = 5$ cm, and supposing both measurements have a precision of ± 0.1 cm,

   (a) What are the bounds of uncertainty on calculated values of the surface area and volume of the cylinder?

   (b) Calculate the sensitivities of the equations for the surface area and volume of the cylinder. To which variable are the surface area and volume calculations more sensitive: h or r? What about π?

6. Calculate the sensitivity of θ to small errors of ϵ for the Lake Las Vegas water quality model for generic values of ϵ_0 and θ_0. The Lake Las Vegas model is given by

$$\theta = \frac{1 - e^{-\epsilon\tau} + \epsilon e^{-\epsilon\tau}}{\epsilon}. \tag{10.51}$$

   (a) Does the sensitivity increase, decrease, or stay the same with increasing time (τ)?

   (b) Suppose $\epsilon_0 = 0.5$ and $\theta_0 = 1.5$. What is the sensitivity?

   (c) What is the sensitivity of $\theta(\tau \to \infty)$ to small errors in ϵ? How does this value compare with the sensitivities found in the previous parts of this question?

Notes

1 Of course, these days most people would just pick up a calculator and punch a few buttons. But this is an example of a general principle, rather than an actual problem to solve.

2 In my classes, I often refer to the Taylor series as "the series that shall not be named," because students that have encountered it in one of their calculus classes often have an almost pathological fear of it.

3 An analytic function is one in which the real and imaginary parts of the function, designated by the letters u and v, respectively, satisfy the Cauchy–Riemann equations,

$$\frac{\partial u}{\partial x} = \frac{\partial v}{\partial y}$$

and

$$\frac{\partial u}{\partial y} = -\frac{\partial v}{\partial x}.$$

Any analytic function has a Taylor series representation.

4 With the restriction that the approximation only holds within the radius of convergence to the nearest singularity.

5 The use of series representations to simplify complex equations and to exploit small parameters to develop solutions is known as "perturbation methods." For those readers who, like myself, find this topic endlessly fascinating, I highly recommend the well-written and clear presentation of the book by Nayfeh (1981).

6 Much of the following discussion is based on the excellent article "Robustness–the ups and downs of ancient astronomy" in Meyer (2004).

7 The accuracy of a measurement is defined as the measurement's closeness to the true value. The precision of a measurement is a function of how repeatable it is, that is, the number of significant figures. In general, accuracy (or lack thereof) is a function of systematic error, whereas precision is associated with random errors.

8 There may be other, theoretical, applications of such a great number of significant figures for π of which I am unaware. For most practical applications, however, equations usually show a sensitivity on the order of 1 to the value of π. Most investigators would be better off spending their efforts on getting more careful measurements of the radii, the angles, or whatever other quantities parameterize their models, and are invariably more difficult to evaluate precisely than the value of π.

References

Meyer, W.J. (2004) *Concepts of Mathematical Modeling*, Dover Publications, Inc., Mineola.
Nayfeh, A.H. (1981) *Introduction to Perturbation Techniques*, John Wiley & Sons, Inc., New York.

CHAPTER 11

A case study

Chapter summary

In the foregoing chapters, we have looked over the main equations that are used in groundwater hydrology, examined some methods for their solution, and thought about how to translate a conceptual understanding of a given system into a quantitative expression. Although the systems we examined were more or less idealized, the approaches used were designed to apply to real-life problems—not just mythical landscapes. In this final chapter, we will attempt to synthesize everything we have learned and apply it to the analysis of a real-life site: the Borax Lake hot springs (BLHS) of southeast Oregon, USA. The model we will develop is simple, but it gives insight into the functioning of the system (as well as into the construction of hydrologic models), and serves as a jumping-off point for more sophisticated modeling and analysis.

11.1 The Borax Lake Hot Springs

The central and southeast portion of the state of Oregon, USA, is a sprawling, semiarid to arid region known as the Oregon high desert. The area has an average elevation of about 1200 masl, and averages about 380 mm/year of rainfall. Much of the Oregon high desert is also part of the US Basin and Range province, which is a vast area of horsts and grabens that can possibly be attributed to accommodation of the strike-slip motion of the San Andreas fault by east–west trending extension and crustal thinning. Heat flow in the Basin and Range is variable but generally high, averaging around 85 mW/m² and ranging from a low of about 60 mW/m² to >100 mW/m² (Blackwell, 1983). High heat flow in the region is responsible for the concentration of geothermal resources in the area, and has given rise to a large number of hot springs that are an important part of the regional ecology, in addition to being well known to recreationalists.

One of the hydrothermal areas in southeast Oregon that is of special interest to me is the BLHS site, in the Alvord Basin of Harney County, Oregon. The Borax Lake area is somewhat unusual, consisting of about 175 hot springs, linearly organized over an 800 m distance along the trace of the Borax Lake

Models and Modeling: An Introduction for Earth and Environmental Scientists, First Edition. Jerry P. Fairley.

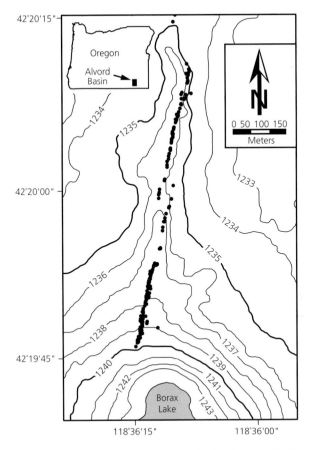

Figure 11.1 Study area location and site map, showing the distribution of springs (indicated by black dots) in the vicinity of Borax Lake, with elevations given in meters above sea level. Figure taken from Fairley (2009). © John Wiley & Sons.

fault (Figure 11.1). The springs range in temperature from around 21 °C to as much as 94 °C (Fairley and Hinds, 2004b), which is a few degrees below the nominal boiling point of water at the altitude of the site. When I first visited Borax Lake, I became convinced that the physical attributes of the springs contained information about the hydrological characteristics of the controlling fault. Ultimately, my students, collaborators, and I published more than a half-dozen peer-reviewed articles on the Borax Lake springs, most of which were focused on understanding the permeability of the fault and the way in which it controlled hydrothermal discharge at the site (Fairley et al., 2003; Fairley and Hinds, 2004a, 2004b; Heffner and Fairley, 2006; Fairley and Nicholson, 2006; Hess et al., 2009; Fairley, 2009). For additional background on the geology and hydrology of the BLHS site, I recommend Heffner and Fairley (2006).

11.2 Study motivation and conceptual model

The Borax Lake site is unusual for several reasons, foremost of which is the number of hot springs that discharge along the strike of the fault. Perhaps even more intriguing is the extreme variability in the temperatures of the water discharging from the springs, where two springs, 1 or 2 m apart, can differ in temperature by as much as 30 °C (Fairley et al., 2003). Originally, we formed the hypothesis that the temperature variations resulted from variable amounts of mixing between cooler, shallow groundwaters and higher temperature water coming from depth. However, geochemical samples from the springs show no evidence of a mixing trend.[1] Alternatively, the hypothesis was formed that the observed temperature variations are the result of different amounts of cooling along the individual flowpaths from depth to the discharge points at the land surface. Presumably, different amounts of cooling are reflective of the effective permeabilities of the flowpaths, with flowpaths having lower permeability allowing greater cooling during ascent. Thus, the distribution of temperatures observed in the springs may be a function of the permeability distribution of the fault.

Because the surface expression of the discharge area is linear and associated with the plane of the controlling fault, it appears likely that a model of the springs can be constructed in two dimensions (x, z), at least until it can be shown that 3D simulations are required to reproduce the behavior of interest. On the other hand, the strong heterogeneity that apparently controls the behavior observed at the site introduces a factor that might be difficult to incorporate into an analytical model. However, before moving to complex numerical simulations, it would be useful and instructive to have a simple but quantitative model of the site to guide expectations for more comprehensive models. Is there some appropriately simplified conceptual model that would represent the broad-brush behavior of the site, while still remaining analytically tractable?

What comes to mind in this case is to examine each spring individually as a 1D "straw," a tube or pipe with effective hydrologic characteristics that are uniform for that particular tube, but which characteristics could represent the flowpath for any of the individual springs at BLHS. Water is assumed to start from some reservoir at depth, which is approximately the same temperature for any of the springs at the site. The fluids rise through the tube under the influence of the hydraulic gradient at a rate determined by the hydraulic characteristics of the tube (permeability, cross sectional area, etc.), losing heat at some rate along the flowpath, and ultimately discharge at the land surface. Although undoubtedly an extremely simplified picture of the actual site, the model roughly captures the essence of our conceptual understanding of the factors controlling discharge temperature at the BLHS site, and may be simple enough to solve analytically. Even if, in the end, the model is either too difficult to solve analytically or too simple to represent the observed behavior, at the very least the time and

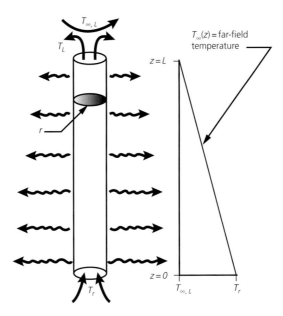

Figure 11.2 Schematic diagram showing the conceptual model for the Borax Lake hot springs problem. Figure taken from Fairley (2009). © John Wiley & Sons.

effort spent in developing it will help us to better understand the site hydrology. A simple schematic of the conceptual model is presented in Figure 11.2.

11.3 Defining the conceptual model

In thinking about the problem as described in the earlier paragraphs, we have begun to develop a conceptual understanding of how we might represent the system. However, the alert reader will have noted that, to this point in our discussion, we haven't defined the actual purpose of our model, outlined what we know and what we assume about our system, and so on. The process of developing a model is not a static, step-by-step affair but, at some point early in any investigation, there needs to be a clear definition of the problem—as has been pointed out numerous times in the preceding chapters. We have reached that point now, and the next step is to refine our somewhat hazy understanding of what we would like to achieve into an exact statement of the purpose of the model.

FIND: A mathematical expression for the effective permeability of the flowpath of a given spring at the BLHS site, as a function of the discharge temperature measured at the land surface.

Note that we have stated succinctly our objective (to find a way to estimate flowpath permeability), and we have tied the achievement of our objective to an observable quantity (the discharge temperature). We have also made it clear that our model only has to apply to one spring at a time, and we have stated that an *effective* value of permeability will be an acceptable outcome—we don't have to know the permeability structure of the flowpath in detail (although we will have more to say about this in the final paragraphs of this chapter).

The reader will know, of course, that next comes a statement of what is known about the site.

KNOWN:

- The discharge temperatures, elevations, and GPS locations (sub-meter precision) of 175 springs at the BLHS site (Heffner and Fairley, 2006).
- pH and major cation hydrochemistries are known for all of the springs (Fairley and Nicholson, 2006); anions are available for a dozen or so (Koski and Wood, 2004), and tritium analyses exist for four springs (Fairley and Nicholson, 2006). It would be extremely helpful to have discharge rates for some or all of the springs in addition to the hydrochemical analyses; unfortunately, only a small fraction of the discharge leaves the springs as surface water. The majority of discharge migrates away from the site in the shallow subsurface, and is therefore not amenable to measurement (Fairley, 2009).
- The temperature of the reservoir is known from geothermometry to be on the order of 200 °C (different geothermometers give temperatures ranging from about 150 °C to 250 °C, Cummings et al. (1993); Koski and Wood (2004)).
- The approximate atmospheric temperatures, precipitation, and so on, are known from a small weather station located in Fields, Oregon (about 8 km distant from the site).
- We also know quite a bit about the local and regional geology, both from the literature (Williams and Compton, 1953) and from site-specific geophysical studies (Hess et al., 2009). For an overview, see Heffner and Fairley (2006).

Next is a list of assumptions.

ASSUME:

- No interaction (mixing or heat exchange) between flowpaths. This probably isn't true, but it greatly simplifies the model to examine only one spring at a time. Moreover, other studies have noted on the basis of geochemical observations that fault controlled springs appear to have limited interactions (Rowland et al., 2008).
- One-dimensional (vertical) flow. Although the flow is likely to be at least 2D (in the plane of the fault), the main vector of flux is in the vertical upward direction, so this may not be a bad assumption. In any case, if the flowpaths have limited interactions (as in the previous assumption), then even if a particular flowpath is tortuous (i.e., not straight), it can still be considered 1D,

although the length will be longer than the straight-line distance between the reservoir and the surface.

- Steady-state flow along a flowpath of constant cross-sectional area. Again, an idealization, but since we don't have any information on what the actual cross-sectional areas of the flowpaths are (or how they vary between the reservoir and the landsurface), it's difficult to see what we could do that is any better.

- The effects of changing density and viscosity as a function of temperature can be neglected. This is a big assumption—it amounts to neglecting buoyant forces, among other things—but it is likely necessary in order to make the problem analytically tractable. If we were only worried about the flow problem, we could perhaps manage to include temperature-dependent density and viscosity, but this is going to be a (weakly) coupled mass-heat transfer problem, and the feedback between the equations in the case of temperature-dependent properties would require a numerical solution. It also means we may as well use the simpler head/hydraulic conductivity representation of potential flow, rather than the more usual (for multiphase or non-isothermal systems) pressure/permeability representation.[2] So, for the time being we will accept this assumption. We can *partially* allow for this by using the average temperature of water along the flowpath when estimating permeability from hydraulic conductivity.

- The temperature boundary at the land surface can be assumed to be constant and equal to the average annual atmospheric temperature. A time-varying temperature upper boundary condition could certainly be incorporated into a numerical simulation, and possibly even into an analytical model, but at least to start we want to keep things reasonably simple.

As we know, we may have to add to (or be able to subtract from) these assumptions as we go along, but this list is a good starting point.

Finally, we have arrived at the point where we can clearly state our strategy for developing our model.

APPROACH: Use mass balance to develop and solve a 1D, steady-state flow model, where hydrothermal fluids flow from the reservoir to the land surface. Use the result (fluid flux or velocity) from the flow model as input to a 1D heat transport model (the ADE), where heat is transported along the flowpath but also lost to the surroundings as the fluid rises. If we can solve the ADE, it should give us a relationship between the discharge temperature of a spring and the hydraulic characteristics of the flowpath feeding that spring.

11.4 Model development

To carry out our plan, we need to develop an expression for mass transport between the reservoir and the surface, as well as an expression that describes heat transport resulting from advection and dispersion of the carrier fluid. We begin with the mass transport equation.

11.4.1 Fluid flow

As should be familiar to you by this time, we will use mass balance to develop a 1D, steady flow model. Using the CV shown in Figure 11.3 to represent the problem, the statement of mass balance is given by

$$M_{in} = M_{out} + \Delta M. \tag{11.1}$$

Assuming the cross-sectional area of the flowpath is circular with radius r, we can expand the components of the mass balance as:

$$M_{in} = q_z \pi r^2 \rho \Delta t, \tag{11.2}$$

$$M_{out} = q_{z+\Delta z} \pi r^2 \rho \Delta t, \tag{11.3}$$

$$\Delta M = 0. \tag{11.4}$$

Equation 11.4, of course, is a reflection of the assumed steady-state nature of the system. Substituting into Equation 11.1, dividing both sides by Δt and Δz, and rearranging, we get

$$\frac{q_{z+\Delta z} - q_z}{\Delta z} = 0. \tag{11.5}$$

Taking the limit as $\Delta z \to 0$ yields

$$\frac{dq}{dz} = 0. \tag{11.6}$$

The next step is to substitute Darcy's law for q; since we assume the hydraulic conductivity to be adequately represented by a single, effective value, we can remove it from under the differentiation and divide both sides, with the result being

$$\frac{d^2 H}{dz^2} = 0. \tag{11.7}$$

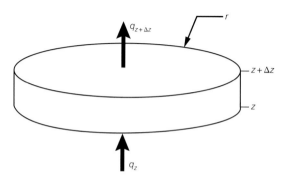

Figure 11.3 Diagram showing the control volume for a flowpath in the Borax Lake problem.

Assuming a flowpath of length L, with the reservoir at $z = 0$ and the discharge point at $z = L$, we have boundary conditions:

$$H(z = 0) = H_0, \tag{11.8}$$

$$H(z = L) = H_L. \tag{11.9}$$

11.4.1.1 Nondimensionalization

We can, of course, nondimensionalize Equations 11.7–11.9 by defining the dimensionless variables:

$$\theta_H = \frac{H - H_L}{H_0 - H_L}, \tag{11.10}$$

$$\xi = \frac{z}{L}. \tag{11.11}$$

We are using the H subscript on the θ variable to distinguish it later from the dimensionless temperature. Substituting into the dimensional equations and boundary conditions gives

$$\frac{d^2\theta_H}{d\xi^2} = 0, \tag{11.12}$$

$$\theta_H(\xi = 0) = 1, \tag{11.13}$$

$$\theta_H(\xi = 1) = 0. \tag{11.14}$$

The reader should be able to verify all the preceding steps.

Equation 11.12 and its boundary conditions should be immediately recognizable as the "Elysian Fields" problem; as a result, we can write the solution directly as

$$\theta_H = 1 - \xi. \tag{11.15}$$

We also know that the dimensionless flux, v, will be equal to 1, and the dimensionless velocity will be the negative of the nondimensional flux divided by the porosity, ϕ,

$$\frac{v}{\phi} = \frac{-1}{\phi} \frac{d\theta_H}{d\xi} = \frac{1}{\phi}. \tag{11.16}$$

We now have all we need, for the time being, from the mass transport portion of the model. Next, we move to the heat transport equation.

11.4.2 Heat transport

It turns out that the energy transport part of the model is going to require more thought than the mass transport, and the making of some tricky decisions. The most difficult question to answer is, how will the loss of heat to the surroundings be represented in the model? This is an important question that deserves careful thought, at least in part because the answer is not immediately obvious.

If we can assume the fluids in the model flowpath are well mixed in a differentially small CV between any z and $z + \Delta z$, the temperature at an arbitrary point along the flowpath is the dependent variable $T(z)$. Likewise, we will assume the temperature far from the flowpath to which the flow is cooling, $T_\infty(z)$, (i.e., the "far-field temperature") is a linear function of depth[3] as

$$T_\infty(z) = T_0 - \frac{T_0 - T_{\infty,L}}{L} z, \qquad (11.17)$$

where T_0 is the reservoir temperature and $T_{\infty,L}$ is the average annual air temperature (the ambient land surface temperature). The question is therefore, how does the fluid cool toward the far-field temperature? One possibility is that the far-field temperature exists at some finite distance from the flowpath, in which case the flowpath loses heat according to Poisson's equation, either radial coordinates (for an isolated cylindrical flowpath) or Cartesian coordinates (for a slab-like tabular body of flowpaths coinciding with the plane of the fault). These equations are readily solved (see the exercises at the end of this chapter), but it is not clear at what distance from the flowpath(s) the far-field boundary should be placed. Alternatively, the far-field boundary may be at an infinite distance from the flowpath, analogous to the Theis equation. In the case of a boundary at infinite distance, however, no steady state is possible (as was shown in Chapter 7), and the governing equations for heat flow to an infinite distance cannot meet the required boundary conditions (again, this is explored in the exercises at the end of this chapter).

Instead of either of these choices, we will use a simplified heat sink term in which heat loss is proportional to the difference between the far-field temperature and the temperature of the flowpath,

$$q_r = 2\pi r \Delta z h \left[T(z) - T_\infty(z) \right], \qquad (11.18)$$

where h is the proportionality constant, usually known as the *coefficient of convective cooling* $[ML^2/T^3]$. Equation 11.18 is being used here in a somewhat unorthodox manner: the coefficient of convective cooling is usually employed to quantify heat transfer between a fluid and a solid (as it is here), but from the fluid side. Usually, heat transfer on the solid side is quantified with Fourier's law—the heat transfer equivalent of Darcy's law—and the two sides of the equation are matched up at the boundary (this is a *Robin*, or *boundary condition of the third type*, see Section 5.3.3). The reason heat transfer is treated in this fashion (i.e., using Eq. 11.18, often called *Newton's law of convective cooling*) is that there is no well-defined length scale for heat transfer on the fluid side of a fluid/solid boundary. In this case, there is similarly no well-defined length scale for heat transfer—but on the solid side. As a result, we are twisting the definition of the coefficient of convective cooling (and of Newton's law of cooling) to construct a source term that gives the behavior we need to represent in our model. When used in this

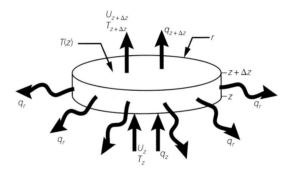

Figure 11.4 Diagram showing the control volume for heat transport (i.e., the energy balance) of the Borax Lake problem.

way, the coefficient of convective cooling is acting in a similar fashion to the "leakance" term that is sometimes used to represent recharge from overlying layers in models of aquifer tests with leaky confining units (e.g., Hantush and Jacob, 1955).

We can now apply an energy balance to the quantification of our conceptual model for heat transport. The balance equation is

$$E_{in} = E_{out} + \Delta E. \tag{11.19}$$

Using the schematic diagram of the CV shown in Figure 11.4, we can expand the terms of Equation 11.19 as

$$E_{in} = U_z T_z \pi r^2 \rho c \Delta t + q_z \pi r^2 \Delta t, \tag{11.20}$$

$$E_{out} = U_{z+\Delta z} T_{z+\Delta z} \pi r^2 \rho c \Delta t + q_{z+\Delta z} \pi r^2 \Delta t$$
$$+ 2\pi r \Delta z h \Delta t \left[T(z) - T_\infty(z)\right], \tag{11.21}$$

$$\Delta E = 0, \tag{11.22}$$

where c is the specific heat of the fluid $[L^2/T^2\Theta]$ (in SI units, e.g., J/kg °C) and $T_\infty(z)$ is given by Equation 11.17. The $\Delta E = 0$ term is, of course, a consequence of the assumed steady-state nature of the problem. Substituting and rearranging yields

$$\pi r^2 \rho c \Delta t \left[U_{z+\Delta z} T_{z+\Delta z} - U_z T_z\right] + \pi r^2 \Delta t \left[q_{z+\Delta z} - q_z\right]$$
$$= -2\pi r h \Delta z \Delta t \left[T(z) - T_\infty(z)\right], \tag{11.23}$$

$$\left[U_{z+\Delta z} T_{z+\Delta z} - U_z T_z\right] + \frac{1}{\rho c} \left[q_{z+\Delta z} - q_z\right]$$
$$= -\frac{2h\Delta z}{r\rho c} \left[T(z) - T_\infty(z)\right]. \tag{11.24}$$

Dividing both sides by Δz and taking the limit as $\Delta z \to 0$:

$$\frac{d[UT]}{dz} + \frac{1}{\rho c} \frac{dq_z}{dz} = -\frac{2h}{r\rho c} [T(z) - T_\infty(z)].$$ (11.25)

As discussed before, the analog to Darcy's law for heat transfer is called Fourier's law,

$$q = -K_T \frac{dT}{dz},$$ (11.26)

where K_T is the *thermal conductivity* (analogous to the hydraulic conductivity in Darcy's law). Substituting into Equation 11.25 and assuming K_T is a constant, we have

$$\frac{d[UT]}{dz} - \frac{K_T}{\rho c} \frac{d^2 T}{dz^2} = -\frac{2h}{r\rho c} [T(z) - T_\infty(z)].$$ (11.27)

In Equation 11.27, the parameter group,

$$D = \frac{K_T}{\rho c},$$ (11.28)

is called the *thermal diffusivity* $[L^2/T]$.[4] In addition, we know from our work in Section 11.4.1.1 that U, the advective fluid velocity of a flowpath, is a constant, so we can remove it from under the differential, rearranging to obtain

$$\frac{d^2 T}{dz^2} - \frac{U}{D} \frac{dT}{dz} = \frac{2h}{rK_T} [T - T_\infty(z)].$$ (11.29)

You should recognize this as the ADE from Chapter 8, specialized for steady and uniform flow, and including a source term (the heat loss term on the right-hand side).

Because Equation 11.29 is a second-order ODE, it should have two associated boundary conditions. These could be specified as a type 1, 2, or 3 condition at each of the top and bottom boundaries of the model (i.e., at the reservoir and the land surface), or possibly as Cauchy-type conditions at the lower boundary (i.e., at the reservoir). For the time being, we will specify a type 1 condition on the lower boundary:

$$T(z = 0) = T_0,$$ (11.30)

and leave the upper boundary condition to be specified later if required.

11.4.2.1 Nondimensionalization

As we have done throughout this text, we begin the process of nondimension-alization of Equation 11.30 by defining a dimensionless temperature, θ_T, and a dimensionless length, ξ:

$$\theta_T = \frac{T - T_{\infty,L}}{T_0 - T_{\infty,L}}, \tag{11.31}$$

$$\xi = \frac{z}{L}, \tag{11.32}$$

where $T_{\infty,L}$ is the average annual temperature at the land surface, T_0 is the temperature at depth (at the reservoir), and L is the length of the flowpath (the same as for the flow equation in the previous section). Substituting into Equation 11.29 yields

$$\frac{d^2\theta_T}{d\xi^2} - \left(\frac{UL}{D}\right)\frac{d\theta_T}{d\xi} = \left(\frac{2hL^2}{rK_T}\right)[\theta_T - (1-\xi)]. \tag{11.33}$$

We can write Equation 11.33 in a more useful (for our present purposes) form by dividing through by UL/D to arrive at

$$\left(\frac{D}{UL}\right)\frac{d^2\theta_T}{d\xi^2} - \frac{d\theta_T}{d\xi} = \left(\frac{2hL}{Ur\rho c}\right)[\theta_T - (1-\xi)]. \tag{11.34}$$

Here, we have simplified the dimensionless parameter on the right-hand side of the equation by decomposing the diffusivity D into its component parameters and canceling terms.

There are two dimensionless groups of parameters in Equation 11.33; the first of these we can recognize from Chapter 8 as the Peclet number,

$$\lambda = \frac{UL}{D}, \tag{11.35}$$

which, you will recall, quantifies the relative importance of advective transport to diffusive transport. The second is a new dimensionless parameter we will call β that quantifies the strength of the sink term (i.e., how rapidly heat is lost from the flowpath to the surroundings),

$$\beta = \frac{2hL}{Ur\rho c}. \tag{11.36}$$

The final form of the dimensionless governing equation is therefore

$$\frac{1}{\lambda}\frac{d^2\theta_T}{d\xi^2} - \frac{d\theta_T}{d\xi} = \beta[\theta_T - (1-\xi)]. \tag{11.37}$$

The dimensionless lower boundary condition is, of course

$$\theta_T(\xi = 0) = 1. \tag{11.38}$$

11.4.3 Solving the equation

The first thing that should come to mind when attempting to solve Equation 11.37 is "would it be possible to simplify the equation, without introducing appreciable error?" In this case, we can look at some approximate values of the Peclet number to see if the diffusion term can perhaps be neglected. The thermal diffusivity of water is on the order of 10^{-7} m^2/s, and the length scale of the problem is going to be somewhere between $L = 100$ m and $L = 1000$ m. At this point we don't know what U is, but we can make a guess that it might be on the order of, say 0.01 m/s (1 cm/s). This gives us a rather large Peclet number (around 10^6, assuming $L = 100$ m). Even if the fluid velocity is several orders of magnitude smaller than this, the Peclet number will be large (i.e., $\lambda > 10$), so we are probably justified in neglecting the diffusion term (the second-order derivative).[5] In this case, our governing equation reduces to

$$\frac{d\theta_T}{d\xi} = -\beta\theta_T + \beta(1 - \xi). \tag{11.39}$$

Can we solve this simplified equation? In fact we can, using an extension of the direct integration technique called *variation of parameters* (Logan, 2006). We first solve the associated homogeneous problem[6]:

$$\frac{d\theta_T}{d\xi} = -\beta\theta_T, \tag{11.40}$$

$$\int \frac{d\theta_T}{\theta_T} = -\beta \int d\xi, \tag{11.41}$$

$$\theta_T = Ce^{-\beta\xi}, \tag{11.42}$$

where the C in Equation 11.42 is the constant of integration. We now assume the constant of integration is going to somehow accommodate the nonhomogeneous part of Equation 11.39; we can show this dependence explicitly by writing:

$$\theta_T = C(\xi)e^{-\beta\xi}. \tag{11.43}$$

We can find the derivative of θ_T,

$$\frac{d\theta_T}{d\xi} = C'e^{-\beta\xi} - \beta Ce^{-\beta\xi}, \tag{11.44}$$

where the prime indicates differentiation with respect to ξ. Substituting Equations 11.43 and 11.44 into Equation 11.39 yields an ordinary differential equation for $C(\xi)$:

$$C'e^{-\beta\xi} - C\beta e^{-\beta\xi} = -C\beta e^{-\beta\xi} + \beta(1 - \xi), \tag{11.45}$$

$$C'e^{-\beta\xi} = \beta(1 - \xi), \tag{11.46}$$

$$\frac{dC}{d\xi} = \beta(1 - \xi)e^{\beta\xi}. \tag{11.47}$$

This last expression can be integrated to find

$$C(\xi) = e^{\beta\xi}\left[1 - \xi + \frac{1}{\beta}\right] + D, \tag{11.48}$$

where D is the constant of integration. The general solution of Equation 11.39 is therefore

$$\theta_T = \left[1 - \xi + \frac{1}{\beta}\right] + De^{-\beta\xi}. \tag{11.49}$$

Setting $\xi = 0$ and $\theta_T = 1$ (from the condition on the lower boundary, Equation 11.38), we find $D = -1/\beta$. The solution of Equation 11.39 is therefore

$$\theta_T = 1 - \xi + \frac{1 - e^{-\beta\xi}}{\beta}. \tag{11.50}$$

Equation 11.50 gives the dimensionless temperature at any point along the flowpath (from $0 \le \xi \le 1$) as a function of the nondimensional distance from the reservoir and the dimensionless heat transfer coefficient β. A plot of the solution for select values of β is shown in Figure 11.5.

Of course, we still need to relate our solution (Equation 11.50) to our observable quantity. By the obvious specialization of setting $\xi = 1$, we can obtain the equation for the discharge temperature of the spring at the land surface, θ_L:

$$\theta_L = \frac{1 - e^{-\beta}}{\beta}. \tag{11.51}$$

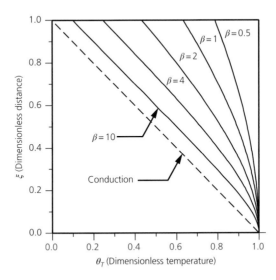

Figure 11.5 Plot of the solution for the dimensionless temperature, θ_T, of a hydrothermal fluid, migrating from a reservoir (at $\xi = 0$) to the land surface (at $\xi = 1$) for a selection of values of the dimensionless heat transfer coefficient β. Figure taken from Fairley (2009). © John Wiley & Sons.

11.4.4 Finding permeability

We are now close to the point at which we originally aimed: an expression to estimate the flowpath permeability from an observation of discharge temperature at the land surface. In fact, Since β is defined as

$$\beta = \frac{2hL}{Ur\rho c},\qquad(11.52)$$

we can rearrange to find the velocity, U, of the fluid rising in the flowpath:

$$U = \frac{2hL}{\beta r\rho c}.\qquad(11.53)$$

Furthermore, we know the velocity is related to the fluid flux by the porosity, ϕ,

$$U = \frac{q}{\phi} = -\frac{K}{\phi}\frac{dH}{dz} = -\frac{k\rho g\,(H_0 - H_L)}{\mu\phi L}\frac{d\theta_H}{d\xi}.\qquad(11.54)$$

We know that $d\theta_H/d\xi = -1$ from Equation 11.16. Substituting and rearranging we have

$$\frac{k\rho g\,(H_0 - H_L)}{\phi L} = \frac{2hL}{\beta r\rho c},\qquad(11.55)$$

$$k = \frac{2h\phi}{\beta r\,(H_0 - H_L)}\left(\frac{L}{\rho c}\right)^2.\qquad(11.56)$$

We know β as a function of the discharge temperature at the land surface from Equation 11.51. Although Equation 11.51 can't be rearranged to give an *explicit* expression for β, it does form an implicit relationship:

$$\beta = \beta\,(\theta_L),\qquad(11.57)$$

Appendix B gives more details on solution methods for implicit equations. In any case, Equation 11.56 supplies the required statement relating permeability of the flowpath to the spring discharge temperature—provided we know the other parameters in the equation. In that sense, we can say that we have achieved the goal we set for ourselves in our "find" statement.

11.5 Evaluating the solution

Although I haven't named it explicitly as I have the other statements, there is an additional part of solving a problem that goes along with the "find," "known," "assume," and "approach" statements: the "comments" section. In the format of this text, the comments on the problem have been somewhat subsumed into the general discussion; however, there comes a point in any problem-solving endeavor when the original goal of the study has been attained—but the work isn't yet done. Of course it will be necessary to write a paper or report on your investigation; but, more fundamentally, once the goal is achieved it is time to stop and consider the solution for a moment. Admire the solution a little bit. Play with it. Ask yourself some questions about it: can it be extended or generalized? Are there useful specializations that can be made? What happens if the independent variable becomes larger or smaller than the intended range—are the results physically meaningful? What if the independent variable becomes negative? What can we learn from this equation about other parameters in the system, or what other systems can we apply this equation to? What are the sensitivities of the equation, and what parameters are known or unknown? These questions go on and on, and it is important to spend time considering them. In what follows, I will give an example of these types of "comments," but you shouldn't take my comments on this particular problem as a template for your own problems. You undoubtedly expended considerable time and thought on finding a solution to your problem—spend some time revelling in it.

11.5.1 Estimating permeability

The first thing we should do, of course, is to use our new equation to make a rough estimate of permeability at the Borax Lake site. The student will have the opportunity to make a more refined estimate in the exercises; here, we will concentrate on making order of magnitude estimates to get a general idea of the range of permeabilities we are likely to encounter at the site.

For a more careful analysis, we would probably want to calculate values for the density and viscosity of water at the average temperature along the flowpath. However, since we are making an order of magnitude estimate here, we will make do with the following parameter estimates (SI units): $\phi_0 = 10^{-1}$, $h = 10^1$ W/m^2°C, $\beta = 10^0$, $r = 10^0$ m, $\rho = 10^3$ kg/m^3, and $c = 10^3$ J/kg°C. The length of flowpath, L, is probably the most uncertain parameter, and it may vary by an order of magnitude between $L = 10^2$ and $L = 10^3$ m; we can also assume that ΔH will be of the same order of magnitude as L, varying from $10^2 \leq (H_0 - H_L) \leq 10^3$ m, according to the value of L.[7] On the basis of these numbers, we get a minimum permeability value of $k = 10^{-10}$ m^2 (assuming $L = 10^2$ m), and a maximum permeability of $k = 10^{-9}$ m^2 (assuming $L = 10^3$ m). You may want to check the order of magnitude parameter estimates and try to figure out how I came up with them, how I used them to calculate k, and decide whether or not you agree with them.

Now that we have an estimated range of permeability, we should also check the Peclet number of the transport equation, to see if we were justified in neglecting the diffusion (conduction) term. Recall that the Peclet number is given by

$$\lambda = \frac{UL}{D}. \tag{11.58}$$

Since we can estimate U from the linearized form of Darcy's law,

$$U = -\frac{K}{\phi}\frac{H_L - H_0}{L} = \frac{k\rho g}{\phi\mu}\frac{H_0 - H_L}{L}, \tag{11.59}$$

we can re-write the Peclet number as

$$\lambda = \frac{k\rho g\,(H_0 - H_L)}{\phi\mu D}. \tag{11.60}$$

Depending on whether we use $L = 10^2\,\text{m}$ or $L = 10^3\,\text{m}$, and applying our order of magnitude estimates of the other parameters (including our estimates of k as appropriate for the value of L), we get values of the Peclet number from order of 10 to 100. So, we are a little close at the lower end, but we're okay on the high end. For the level of simplification of our model, then, our decision to drop the diffusion term was probably a good one.

11.5.2 Checking the limits

We should, of course, check the behavior of the solution in the limits to make sure it acts as we expect. It is easy to check that the solution meets the lower boundary condition when $\xi = 0$, and when $\xi = 1$ the solution reduces to Equation 11.51. What happens when β goes to the limits of its range, though?

We can expect that β would be restricted to a range between $0 < \beta < \infty$, based on physical arguments. When $\beta \to \infty$ we have

$$\lim_{\beta\to\infty} \theta_L \to \frac{1}{\infty} \to 0, \tag{11.61}$$

which is the conduction case (i.e., the discharge temperature is the same as the annual average air temperature). On the other hand, when β goes to 0, we can apply L'Hopital's rule to find

$$\lim_{\beta\to 0} \theta_L \to 1, \tag{11.62}$$

which is the same as saying the water comes up so quickly it doesn't have time to cool (you should verify this limit for yourself). Both these cases therefore make sense. But what is the actual range of β we observe at the Borax Lake site?

The maximum and minimum spring temperatures observed at the Borax Lake site are approximately 21 and 94 °C (Fairley, 2009). If we assume a reservoir temperature of 250 °C, and an average annual air temperature of 5 °C, we can calculate the minimum and maximum dimensionless discharge temperatures to be $\theta_L = 0.065$ and $\theta_L = 0.363$, respectively. These discharge temperatures equate to values of $\beta(\theta_L = 0.065) = 15.38$ and $\beta(\theta_L = 0.363) = 5.54$, which is therefore the range of β values that should exist at the site. Discharge temperatures much higher than the highest observed temperature would result in boiling, which would invalidate the model used here (because the model does not account for multiphase heat transfer). Lower discharge temperatures (and thus higher β

values) certainly exist at the site, but the permeability of the conduits is probably too small to permit sufficient discharge to form springs.

11.5.3 What permeability is calculated?

Perhaps the next question we should ask is as follows: "what is the physical significance of the permeability we have calculated?" With any model, there is the danger of drifting off into a non-physical "la-la land." It should certainly be clear that any model that gives a single value of a parameter (i.e., permeability) for an entire system (a given flowpath) should be examined closely.

We can perform a basic "thought experiment" to investigate this question. Suppose we were to pack a tube full of layers of material of different permeabilities. That is, if we had a vertical tube, 10 m long, and in the bottom 1/2 m we packed some coarse sand, then in the next 1/2 m we packed some silt, then some fine gravel in the next 1/2 m, and so on until the tube is filled to the top. Next, we run water up the tube, with the water starting off at some temperature at the base, and we measure the temperature of the water discharging from the top. If we then apply our model to calculate an "effective permeability," how would that calculated value relate to the permeabilities of the system?

If you look back at our discussion of flow across a layered medium in Section 9.3.1.2, you can see that the effective permeability of the tube would be calculated using *harmonic averaging* (Eqs. 9.24 and 9.25). Harmonic averaging has an important property that is relevant here: *harmonic averaging most heavily weights the smallest value of the quantity being averaged.* The harmonic average will be very close to the magnitude of the lowest permeability layer. Another way to say this is the lowest permeability region along the flowpath will dominate the permeability of the flowpath as a whole. The result, then, is that the permeability we calculate with Equation 11.56 will be *a minimum* for the flowpath—it represents the most restricted portion of the flowpath, which exerts overall control on the flow in our conceptual tube.

11.5.4 What are the model sensitivities?

We should certainly take the time to estimate the sensitivity of our model to small errors of the parameters. As an example, we can quickly find the sensitivity of our estimate of k to small errors in ϕ,

$$S(k, \phi) = \left(\frac{\phi_0}{k_0}\right)\frac{dk}{d\phi}(\phi = \phi_0), \tag{11.63}$$

$$S(k, \phi) = \left(\frac{\phi_0}{k_0}\right)\frac{2hL^2}{\beta r\,(H_0 - H_L)\,\rho^2 c^2}\frac{d\phi}{d\phi}, \tag{11.64}$$

$$S(k, \phi) = \left(\frac{\phi_0}{k_0}\right)\frac{2h}{\beta r\,(H_0 - H_L)}\left(\frac{L}{\rho c}\right)^2. \tag{11.65}$$

At this point we don't have precise values for most of the parameters in Equation 11.65. However, if we substitute our expression for k_0 the terms cancel, and we get a sensitivity around $S(k, \phi) = 1$, which means the model is not overly sensitive to small errors in porosity.

We can also try to calculate the sensitivity of our estimate of k to small errors in L.

$$S(k, L) = \left(\frac{L_0}{k_0}\right) \frac{dk}{dL} (L = L_0),\tag{11.66}$$

$$S(k, L) = \left(\frac{L_0}{k_0}\right) \frac{2h\phi}{\beta r (\rho c)^2 (H_0 - H_L)} \frac{d\left[L^2\right]}{dL} (L = L_0),\tag{11.67}$$

$$S(k, L) = \frac{4h\phi L_0^2}{k_0 \beta r (\rho c)^2 (H_0 - H_L)}.\tag{11.68}$$

Again, we can substitute our expression for k_0 and cancel parameters; this time, we get a sensitivity of about 2—twice that for porosity. Our expression for permeability is more sensitive to errors in the length of flowpath than porosity, and a (say) 5% relative error in our estimate of the length of the flowpath will translate into a 10% error in the estimated permeability.

The last sensitivity we will estimate is the sensitivity of our estimate of k to small errors in our observed quantity, θ_L. This is a more difficult calculation than the others, because θ_L enters our expression for k through our estimate of β. Using the chain rule, we have

$$S(k, \theta_L) = \left(\frac{\theta_{L,0}}{k_0}\right) \frac{dk}{d\theta_L} = \left(\frac{\theta_{L,0}}{k_0}\right) \frac{dk}{d\beta} \frac{d\beta}{d\theta_L}.\tag{11.69}$$

The derivative of k with respect to β is easy:

$$\frac{dk}{d\beta} = \frac{2h\phi}{r (H_0 - H_L)} \left(\frac{L}{\rho c}\right)^2 \left(\frac{-1}{\beta^2}\right),\tag{11.70}$$

$$\frac{dk}{d\beta} = -\frac{2h\phi}{r (H_0 - H_L)} \left(\frac{L}{\rho c \beta}\right)^2.\tag{11.71}$$

The derivative of β with respect to θ_L, however, is more difficult, because we need to use implicit differentiation. Taking the derivative of both sides of Equation 11.51 and using the chain rule on the terms that include β:

$$\frac{d\theta_L}{d\theta_L} = \frac{d}{d\theta_L} \left[\frac{1}{\beta} + \frac{e^{-\beta}}{\beta}\right],\tag{11.72}$$

$$1 = -\frac{\beta'}{\beta^2} + \left[\frac{-\beta' e^{-\beta}}{\beta^2} - \frac{\beta' e^{-\beta}}{\beta}\right],\tag{11.73}$$

$$1 = -\beta' \left[\frac{1 + e^{-\beta} + \beta e^{-\beta}}{\beta^2}\right],\tag{11.74}$$

$$\frac{d\beta}{d\theta_L} = \frac{\beta^2}{(1 + \beta) e^{-\beta} - 1}.\tag{11.75}$$

where the primes indicate differentiation with respect to θ_L. Putting all the pieces together, we have

$$S(k, \theta_L) = \left(\frac{\theta_{L,0}}{k_0}\right) \frac{-2h\phi}{r(H_0 - H_L)\left[(1 + \beta)e^{-\beta} - 1\right]} \left(\frac{L}{\rho c}\right)^2. \tag{11.76}$$

Although we can substitute into Equation 11.76 for k_0, we can't readily substitute in for β because we have an implicit relationship between β and θ_L. However, we can substitute $\theta_L(\beta)$ for $\theta_{L,0}$ and obtain

$$S(k, \theta_L) = \frac{1 - e^{-\beta}}{1 - (1 + \beta)e^{-\beta}}. \tag{11.77}$$

As β varies over a range of $1 \leq \beta \leq 10$, the sensitivity goes from 2.4 to 1.0005, respectively. For the range of θ_L observed on the site, we have sensitivities going from $S(k, \theta_L = 0.065) = 1$ ($\beta = 15.38$) to $S(k, \theta_L = 5.54) = 1.02$ ($\beta = 5.54$), which is to say that the model is not particularly sensitive to small errors in estimates of the observed discharge temperature.

11.6 Conclusions

In this chapter, we have constructed a model of fluid flow and heat transport that can be applied to individual springs at the Borax Lake hot springs site in southeast Oregon, USA. The model is a simplified representation of advective heat transport from a hydrothermal reservoir at depth to a discharge point at the land surface, and allows an estimate of flowpath permeability to be made from an observable quantity (the discharge temperature of the spring at the land surface), provided several additional parameters characterizing the system are known or can be estimated.

The model gives some insight to the subsurface hydrologic structure of the Borax Lake fault, but is really only the first step toward understanding the hydraulic architecture of the site. In the original article upon which this chapter is based (Fairley, 2009), the model presented here was used to condition stochastic realizations of the fault permeability structure, and a number of conclusions were drawn about the effective permeability and correlation structure of the fault. Additional inferences have been made regarding the site hydraulic architecture in other, related publications (Fairley and Hinds, 2004a; Hess et al., 2009); the reader with an interest in fault hydrology is referred to those papers for more information.

Although the model presented here is only a preliminary attempt at under-standing a particular site, it makes an excellent example of the approach to mod-eling presented here. By starting simply and making a number of assumptions, we have managed to develop a representation of our site that makes sense, that we can understand, that gives insight, and can be used as a guide to the development of more complex models. It also demonstrates the thought process that a modeler goes through when building a conceptual and mathematical representation of a particular system. In that way, hopefully, it makes a fitting example for this chapter.

11.7 Problems

1. Calculate the sensitivity of the model expression for k to small errors in the estimated value of
   (a) ρ
   (b) $\Delta H = H_L - H_0$
   (c) r
   How do these sensitivities compare with those calculated in the text? What do you think are the least and most sensitive parameters in the model?

2. Although they are only rough approximations, the following formulas can be used to estimate the density and heat capacity of liquid water at various temperatures. For density as a function of temperature,

$$\rho = \frac{\rho_0}{1 + 0.0002 \, (T - T_0)}, \tag{11.78}$$

where T and ρ are the temperature (°C) and density (kg/m^3) of interest, respectively, and T_0 and ρ_0 are the reference temperature and density (you may use $0\,°$C and 999.8 kg/m^3). For heat capacity,

$$c = A + BT + \frac{C}{T^2} + DT^2, \tag{11.79}$$

where T is given in Kelvins (Kelvins $= °$C $+ 273.15$), and the constants are $A = 16.749$, $B = 62.120 \times 10^{-3}$, $C = 32.798 \times 10^5$, and $D = 90.391 \times 10^{-6}$.
   (a) Using Equation 9.26, find average density and heat capacity for fluid along a modeled flowpath at Borax Lake. Use a reservoir temperature of $250\,°$C and a $\beta = 5.54$.
   (b) Using the density and heat capacity calculated in the previous part, what permeability would you expect the flowpath to have? Use a value of $h = 4$ and $L = 1000$ m, and assume ΔH is the same magnitude as L.
   (c) Repeat the previous calculations with $\beta = 15.38$. How different is this permeability from the one you calculated in the previous problem?

3. The following questions are in reference to the problem of quantifying heat flow from a vertical tube (radius r), losing heat to a boundary at temperature T_∞.
   (a) Derive the 1D steady heat flow equation in radial coordinates. For boundary conditions, use $T(r = r_0) = T_{r_0}$ and $T_r(r \to \infty) = T_\infty$.

(b) The steady, 1D radial heat equation is a form of Cauchy-Euler equation, the solution of which is given in any basic textbook on differential equations, or can easily be looked up on the Internet. Find the general solution to the equation. Can you meet the boundary conditions to find the complete solution? If not, why not?

(c) Derive the 1D steady heat flow equation in Cartesian coordinates. For boundary conditions, use $T(x = x_0) = T_{x_0}$ and $T_x(x \rightarrow \infty) = T_\infty$.

(d) Find the general solution for the steady, 1D Cartesian heat flow equation. Can you meet the boundary conditions to find the complete solution? If not, why not?

(e) The Thiem equation is a steady-state solution to fluid flow toward a well that is based on integrating Darcy's law in radial coordinates. The heat transfer equivalent is Fourier's law in radial coordinates.

$$Q = -2\pi r \Delta z K_T \frac{dT}{dr}. \tag{11.80}$$

Solve Fourier's law in radial coordinates to obtain an expression for $T = T(r)$. You may want to look at a derivation of the Thiem equation if you get stuck, but you should try to do this without help first. Could this expression for 1D steady heat flow be used to model heat flow to the surroundings and away from the flowpath in the Borax Lake problem? Why or why not?

Notes

1 There is one exception: a single spring, well off the main trend of the fault, that shows evidence of mixing between the hydrothermal fluids and shallow outflow from nearby Borax Lake (Fairley and Nicholson, 2006).

2 As you are probably aware, hydraulic conductivity (K) is related to permeability (k) by the equation: $K = k\rho g/\mu$. Thus, conductivity is a function of water density (ρ) and viscosity (μ), and indirectly a function of temperature through its effects on density and viscosity.

3 This is a common assumption that supposes the geological materials are of homogeneous thermal conductivity and there is no lateral advective flow.

4 In heat transfer applications, the thermal diffusivity is often represented by the letter α; however, since I like to save the Greek letters whenever possible for dimensionless quantities, I am going to stick to using D for thermal diffusivity, in analogy to other types of diffusivity (e.g., Fickian diffusivity of species, hydraulic diffusivity).

5 Once we solve the simplified equation, we should calculate the actual velocity and recalculate the Peclet number, to be sure we were justified in making this simplification.

6 The homogeneous part of the problem is that portion of the right-hand side that only includes the dependent variable, θ_T, but not the portion that includes the independent variable— in this case, the $\beta(1 - \xi)$ part.

7 If the flowpath is essentially vertical, we know the change in head has to be at least the same magnitude as the length of the flowpath to overcome the elevation difference; otherwise, we wouldn't see any discharge at the surface. If we break out the pressure and elevation head terms separately in the (linearized) gradient, we have

$$\frac{(H_L - H_0)}{L} = \left(\frac{P_L}{\rho g L} + 1\right) - \left(\frac{P_0}{\rho g L}\right) = \frac{P_L - P_0}{\rho g L} + 1. \qquad (11.81)$$

ΔH must be negative for fluid to arrive at the surface; therefore, $(P_L - P_0)/\rho g L$ must be less than or equal to -1 to push the fluids vertically over an elevation gain of L. It will actually have to be a little more negative than -1 to overcome viscous losses; however, for an order of magnitude analysis we can neglect this additional head change.

References

Blackwell, D.D. (1983) Heat flow in the northern Basin and Range province. *The Role of Heat in the Development of Energy and Mineral Resources in the Northern Basin and Range Province*, Special Report 13, Geothermal Resources Council, Dairs, pp. 81–93.

Cummings, M.L., John, A.M.S., and Sturchio, N.C. (1993) Hydrogeochemical characterization of the Alvord Basin geothermal area, Harney County, Oregon, USA. *Proceedings of the 15th New Zealand Geothermal Workshop*, University of Auckland, Auckland, pp. 119–124.

Fairley, J.P. (2009) Modeling fluid flow in a heterogeneous, fault-controlled hydrothermal system. *Geofluids*, **9**, 153–166, doi:10.1111/j.1468–8123.2008.00236.x.

Fairley, J.P. and Hinds, J.J. (2004a) Field observation of fluid circulation patterns in a normal fault system. *Geophysical Research Letters*, **31**, L19502, doi:10.1029/2004GL020812.

Fairley, J.P. and Hinds, J.J. (2004b) Permeability distribution in an active Great Basin fault zone. *Geology*, **32** (9), 825–828.

Fairley, J.P. and Nicholson, K.N. (2006) Imaging lateral groundwater flow in the shallow subsurface. *Journal of Hydrology*, **321** (1–4), 276–285.

Fairley, J.P., Heffner, J., and Hinds, J.J. (2003) Geostatistical evaluation of permeability in an active fault zone. *Geophysical Research Letters*, **30** (18), 1962, doi:10.1029/2003GL018064.

Hantush, M.S. and Jacob, C.E. (1955) Non-steady radial flow in an infinite leaky aquifer. *Eos, Transactions of the American Geophysical Union*, **36** (1), 95–100.

Heffner, J. and Fairley, J.P. (2006) Using surface characteristics to infer the permeability structure of an active fault zone. *Journal of Sedimentary Geology*, **184**, 255–265.

Hess, S., Fairley, J.P., Bradford, J., Lyle, M., and Clement, W. (2009) Evidence for composite hydraulic architecture in an active fault system based on 3D seismic reflection, time-domain electromagnetics and temperature data. *Near Surface Geophysics*, **7**, 341–352.

Koski, A.K. and Wood, S.A. (2004) The geochemistry of geothermal waters in the Alvord Basin, southeastern Oregon. In: Wanty, R.B., and Seal, R.R. (eds.) *Proceedings of the 11th International Symposium on Water-Rock Interaction, Energy and Environment*, Saratoga Springs, NY, June 27 to July 2, 2004, CRC Press, Boca Rat.

Logan, J.D. (2006) *A First Course in Differential Equations*, Springer Science+Business Media, LLC, New York.

Rowland, J.C., Manga, M., and Rose, T.P. (2008) The influence of poorly interconnected fault zone flow paths on spring chemistry. *Geofluids*, **8**, 93–101, doi:10.1111/j.1468–8123.2008.00208.x.

Williams, H. and Compton, R.R. (1953) Quicksilver deposits of Steen Mountain and Pueblo Mountains, southeast Oregon, U.S., *Tech. Rep. Bulletin 995-B*, United States Geological Survey, Reston.

CHAPTER 12

Closing remarks

> Our revels are now ended. These our actors,
> As I foretold you, were all spirits and
> Are melted into air, into thin air:
> And, like the baseless fabric of this vision,
> The cloud capp'd towers, the gorgeous palaces,
> The solemn temples, the great globe itself,
> Yea, all which it inherit, shall dissolve
> And, like this insubstantial pageant faded,
> Leave not a rack behind. We are such stuff
> As dreams are made on, and our little life
> Is rounded with a sleep.
>
> —W. Shakespeare, *The Tempest*

12.1 Some final thoughts

My goal in writing this book was to bring to the forefront some ideas and principles of modeling that are sometimes overshadowed by the glamor of high-end computer simulation. The ability to solve problems involving millions of grid-blocks, with complex boundary conditions, source terms, and variable properties (e.g., permeabilities that vary with saturation and material properties that change in response to applied stress or chemical precipitation and dissolution) means that we are living in a wonderful time for modelers. However, it is easy to overlook the fact that these models rely on basic, underlying mathematical equations. The availability of sophisticated, pre-packaged software is no substitute for conceptual thought or basic analytical skills. I hope I have managed to pass on a little taste for these ideas with this text.

The skillful modeler has a large store of useful tricks, but also a solid background in several areas that are not often found in the standard geology curriculum. In addition to a thorough understanding of basic geology, the best modelers have strong abilities in analytical mathematics, a wide range of computer skills, and a knowledge of statistics. Since most geologists have had little formal coursework in these areas, I would like to take this opportunity to recommend some of the books that have been helpful to me over the years I have been working as a modeler and studying the craft of modeling.

Models and Modeling: An Introduction for Earth and Environmental Scientists, First Edition. Jerry P. Fairley.
© 2017 John Wiley & Sons, Ltd. Published 2017 by John Wiley & Sons, Ltd.
Companion website: www.wiley.com/go/Fairley/Models

Besides a solid grounding in basic calculus (every scientist should have at least two, and preferably three, semesters of calculus), I hope this book has motivated an interest in learning about ordinary and partial differential equations. An excellent book on ordinary differential equations is Logan (2006). Two books I highly recommend for extending the student's abilities in working with partial differential equations are Farlow (1993) and Logan (2004). Another area of mathematics that is of vital importance to modelers (especially those working with numerical analysis and simulation) is linear algebra; an excellent book for this is Kwak and Hong (2004). All of these books are suitable for self-study by the motivated reader.

A good review of basic statistics for geologists, with something of an overview of more advanced topics, is given by McKillup and Dyar (2010). Because hydrogeologists are constantly having to deal with heterogeneous systems, they also have need of some type of spatial statistics; an excellent introduction to the area of geostatistics for hydrogeologists is Kitanidis (1997). For more advanced study of geostatistical techniques, the standard reference on the subject is Rubin (2003).

Every modeler should have the ability to program with greater or lesser facility. As of this writing, the premier scientific computational languages are Fortran, C, or C++, but even facility with a high-level language such as Matlab is a very useful skill for a modeler. If you are primarily interested in using packaged software, then perhaps more important than one of the computational languages is some familiarity with a scripting language such as Perl, Python, or Unix-style shell scripts and commands such as awk, grep, and sed—any of which will help with the preparation of input files and the extraction of information from output files. These types of skills are especially useful when dealing with large datasets. Another useful tool for working on large datasets (or small ones) is R. R is a high-level statistical language with excellent graphical capabilities, similar in form to Matlab. R is open source and freely available on the web, and has been extended with hundreds or thousands of user-written modules to cover a wide range of applications such as time series analysis of data, spatial statistics, coordinate transforms, and so on.

Writing and public speaking are usually not topics covered in hydrology classes, but communication skills are an essential part of modeling, and science in general. Every modeler needs to be able to transmit her/his findings succinctly and clearly to the target audience. Being able to write well and quickly, and to speak smoothly and convincingly, are assets in any career path, but are especially important for scientists and engineers. Professionals that can communicate complex results to clients, regulators, and the general public will always be in demand.

The best modelers have a diverse set of skills and abilities. This book is only a start. Continue to build your toolkit—and, in the meanwhile, read widely, listen carefully, and think deeply!

References

Farlow, S.J. (1993) *Partial Differential Equations for Scientists and Engineers*, Dover Publications, New York.

Kitanidis, P.K. (1997) *Introduction to Geostatistics: Applications in Hydrogeology*, Cambridge University Press, Cambridge.

Kwak, J.H. and Hong, S. (2004) *Linear Algebra*, 2nd edn. Birkhauser, Boston.

Logan, J.D. (2004) *Applied Partial Differential Equations*, 2nd edn. Springer Science+Business Media, LLC, New York.

Logan, J.D. (2006) *A First Course in Differential Equations*, Springer Science+Business Media, LLC, New York.

McKillup, S. and Dyar, M.D. (2010) *Geostatistics Explained: An Introductory Guide for Earth Scientists*, Cambridge University Press, Cambridge.

Rubin, Y. (2003) *Applied Stochastic Hydrogeology*, Oxford University Press, Oxford.

APPENDIX A

A heuristic approach to nondimensionalization

The term "heuristic" refers to an experienced-based approach to problem-solving. When an exhaustive search of all possibilities is impractical, we often resort to heuristic solutions. This is another way of saying that we rely on "common sense," "practical experience," or "rules of thumb."

Finding nondimensional groupings is generally a heuristic endeavor. Engineers (and other students who take engineering classes) will eventually run into the "Buckingham π theorem." The π theorem states that, for an equation with n variables or parameters, described by k fundamental dimensions (e.g., [L], [M], [T], and [Θ] are four dimensions), $p = n - k$ dimensionless parameters can be defined that characterize the relationships between the variables. Unfortunately, the π theorem doesn't tell us what the dimensionless parameters are, guarantee they are physically meaningful, or ensure unique parameter groups are defined. The π theorem is popular in engineering because engineers usually know ahead of time what dimensionless quantities will be involved: if velocity is part of the equation, the Reynolds number will appear; transport equations will include the Peclet number, and so on. In many research situations, however, it is far from clear what dimensionless groups will be important. As a result, in this text we present the following heuristic approach to defining physically relevant dimensionless groups for any given equation.[1]

There are generally three steps to nondimensionalizing an equation, although you often may only need the first two. The steps are as follows:

Step 1: Normalize the variables. You should always do this first. Look in the boundary conditions to find maximum and minimum values of the dependent and independent variables. If a variable has a minimum value different from zero, you can usually subtract the minimum value from the variable, then divide the difference by the range (the maximum minus the minimum values). This will give you a dimensionless variable that ranges from 0 to 1. If the variable naturally ranges from zero to some finite maximum, you can normalize by dividing the dimensional variable by the maximum value. Keep in mind that, when nondimensionalizing the dependent variable, you would like to have the maximum possible number of homogeneous auxiliary conditions (ideally, only one nonhomogeneous condition), so try to find a normalization scheme

Models and Modeling: An Introduction for Earth and Environmental Scientists, First Edition. Jerry P. Fairley.
© 2017 John Wiley & Sons, Ltd. Published 2017 by John Wiley & Sons, Ltd.
Companion website: www.wiley.com/go/Fairley/Models

that maximizes the number of boundaries at which the dependent variable is equal to zero. If, for any of the variables, there is no obvious range (e.g., if the variable ranges from 0 to ∞), define an arbitrary characteristic quantity (e.g., t_c) and divide the dimensionless variable by the arbitrary characteristic quantity. Once you have defined your normalized variables, you should solve for the dimensional variables and their derivatives and substitute back into the governing equation.

Step 2: Simplify the equation. You may be able to cancel terms on both sides of the equation; next, divide both sides by the parameter(s) of greatest magnitude. Check the dimensions of any parameter groupings—they should be dimensionless. At this point, you should define any arbitrary characteristic quantities defined in Step 1 to annihilate the maximum number of parameters. Don't forget to look at the dimensionless boundary conditions, as well, for possible definitions of the characteristic quantities. Whatever parameter groups are left will be the controlling dimensionless parameters of the governing equation. You should examine these groups closely to see if they correspond to known dimensionless numbers, for example, the Fourier number, the Biot number, and the Peclet number.

Step 3: Seek similarity solutions. It sometimes happens during the simplification of the governing equation (Step 2) that the characteristic quantities defined in Step 1 end up "complementing" each other; for example, a characteristic length scale x_c and a characteristic time scale t_c may end up in a dimensionless grouping as x_c^2/Dt_c (where D is diffusivity) or similar. When this happens, it is not possible to simply set the characteristic quantities equal to each other. Instead, try defining a new variable that preserves the apparent relationship between the two characteristic quantities. For the grouping shown before we could try

$$\eta = \frac{x^2}{Dt}.$$ (A.1)

This particular similarity variable is demonstrated in the Theis equation (Chapter 7). Although similarity is not a common situation it does occur, and when it does it is usually possible to use it to achieve substantial simplification of the dimensionless governing equation.

Note

1 The approach to nondimensionalization presented in this appendix and used throughout the text is based on the handout *Nondimensionalization Heuristic* by Professor P.J. Pagni, and the teachings of Professor Pagni in his heat transfer courses at UC Berkeley.

APPENDIX B

Evaluating implicit equations

There is a certain amount of difficulty for some students in understanding the difference between implicit and explicit equations, and in knowing how to evaluate equations of the implicit type. Most people are comfortable with explicit equations (i.e., equations that can be *made* explicit), but implicit equations occur commonly enough in mathematical modeling that an aspiring modeler should have a reasonable understanding of how to approach them.

An *explicit* equation is one in which the independent variables can all be put on one side of the equation, and the unknown can be put on the other side. For example,

$$y = x^2 + 2x + 3, \tag{B.1}$$

$$f(t) = 5e^{-\alpha t}, \tag{B.2}$$

$$f(z) = A\sin(n\pi z), \tag{B.3}$$

In each of these, a value can be substituted into the independent variable on the right-hand side (for x, t, or z) and, by performing the indicated operations, a value can be computed for the dependent variable on the left-hand side (y, $f(t)$, or $f(z)$, respectively).

In contrast, *implicit* equations cannot be solved in a closed-form sense. Take, for example, the implicit equation,

$$\lambda\beta = \tan\beta, \tag{B.4}$$

where λ is a constant, or the equation (Fairley, 2009),

$$\theta_L = \frac{1 - e^{-\beta}}{\beta} \tag{B.5}$$

where θ_L is a known (observed) quantity.

In neither of these equations is there any way to get the desired quantity, β, by itself on one side of the equation so that it may be conveniently calculated. In particular, Equation B.4 comes up quite often in mathematical models of physical systems, and implicit equations in general are encountered often enough that approaches must be found for evaluating them. To the best of my knowledge, there are (at least) four methods for evaluating implicit equations; they are trial-and-error, the graphical method, iteration, and Newton's method. I will present a brief synopsis of each of these approaches in the following text.

Models and Modeling: An Introduction for Earth and Environmental Scientists, First Edition. Jerry P. Fairley.
© 2017 John Wiley & Sons, Ltd. Published 2017 by John Wiley & Sons, Ltd.
Companion website: www.wiley.com/go/Fairley/Models

B.1 Trial and error

Trial and error is probably the most obvious method of solving implicit equations; many of us learned to make a good guess at a value and improve it while trying to shortcut problems in a High School algebra or trigonometry class. However, guessing a number, then making another guess and seeing if the answer is closer than the previous guess, is not a very efficient way of going about finding trial-and-error solutions to equations. Probably the easiest and most straightforward way to apply trial and error to an equation is to first decide in what range of values the answer is expected to lie. Next, decide what precision is required for the application (usually trial-and-error methods work best in situations where the requirements for precision are modest); for example, say you would like the answer to a precision of 0.01. Finally, write a simple computer program (e.g., in Fortran, C++, or Matlab) or use a spreadsheet program to calculate all the values in between the two limits with a step size of 0.01 (or whatever precision you decided on). The answer can then be found by simply reading down through the list of values and selecting the closest one. If the number of values would be too great for such a brute force approach, one can do it in phases, starting with a large range but a coarse step size, then identifying a smaller range and using a finer step size.

Trial and error is not a very elegant way to seek a solution to an implicit equation, but it is simple and understandable. Although the other methods discussed later are more elegant and satisfying (at least to my mind), trial and error is good for quick, one-time estimates, or as a fallback method when all else fails.

B.2 The graphical method

Next to trial and error, the graphical method is probably the oldest approach to finding solutions to implicit equations, and it is generally easy to apply. Unfortunately, the precision that can be obtained from this method is limited; however, it is useful in situations where exact solutions are not required, or as a first step for gaining a good initial guess before refining it with either iteration or Newton's method. Simply stated, the method treats each side of the implicit equation as an explicit equation, and each side is plotted separately, but on the same set of axes. The locations of the intersections of the two plots are the solutions to the complete implicit equation.

For a concrete example, take the following equation:

$$x = 4\pi \cos x. \tag{B.6}$$

Plotting the left-hand side of Equation B.6 gives a straight line running from $y = -4\pi$ (at $x = -4\pi$) to $y = 4\pi$ (at $x = 4\pi$). On the right-hand side is a cosine function with amplitude 4π that runs through four complete cycles on the interval $-4\pi \leq x \leq 4\pi$. When plotted on the same set of axes, the two functions cross at eight points, four negative and four positive values of x, the largest of which is at 4π. (For values of $x < -4\pi$ or $x > 4\pi$, the linear function is less than or greater than, respectively, the amplitude of the cosine function, and no intersections are possible.)

As mentioned before, this method is not terribly precise. It is easy to use, and, besides being intuitive, it has the advantage of drawing the practitioner's attention to equations that have multiple roots. For these reasons, it is a good idea to practice this method and apply it to unfamiliar equations, or in situations where it is not clear what the structure of the solution(s) of a particular equation may be.

B.3 Iteration

Iteration (sometimes known as *fixed-point iteration*, Logan (2006)), is a far more elegant approach to finding solutions to an implicit equation than trial and error. Unfortunately, its success is dependent on way the equation is arranged and (in some cases), on having an appropriate initial guess or estimate of the final answer.

To apply iteration, you should get the unknown alone on one side of the equation (of course, it will also be on the other side of the equation). Using Equation B.5 as an example, we can rearrange slightly to get the following:

$$\beta = \frac{1 - e^{-\beta}}{\theta_L}. \tag{B.7}$$

Now an initial guess is substituted for β on the right-hand side, the indicated operations are performed, and a value for the β on the left-hand side is found. Of course, this second β will not be the same as the initial guess for β (unless the initial guess was extremely lucky indeed). Therefore, we take this new value of β and substitute it on the right-hand side and recalculate the β on the left-hand side. We keep iterating in this fashion until either the difference between two successive iterations drops below a predetermined target precision (i.e., the iteration converges) or we see that the iterations are diverging (getting larger and larger or smaller and smaller without bound). Occasionally, we may find that the solution bounces around without ever converging on any specific value. If the solution fails to converge, we either need to rearrange our equation in a manner more suited to iteration, or use another approach to find a solution. According to Logan (2006), the method converges to a solution x^*, for any initial guess that is "sufficiently close" to the final value, provided the absolute value of the first derivative of the function, evaluated at x^*, is less than 1.

Iteration has the advantage of (usually) being quick to use; in the example given in Equation B.7, six iterations are sufficient to obtain five decimal places of precision (with $\theta_L = 0.2$ and an initial guess of $\beta_0 = 0.5$; the final $\beta_6 = 4.96511$). It is also easy to obtain values to arbitrarily high precision. As noted, its drawbacks are that it doesn't always converge, and the method can be sensitive to the choice of the initial guess. For equations that have multiple roots or even, like Equation B.4, have an infinite number of roots, it may be difficult to find an initial guess that converges to the root of interest. When the method works, however, it is probably the quickest and most accurate of the methods presented here, so it is well worth trying. Sometimes, a few hand calculations will show that the method will work; once this is established, it is easy to throw together a quick-and-dirty computer program if many estimates need to be made.

B.4 Newton's method

Newton's method is another iterative method for finding approximations of the roots of an equation; it is sometimes called Newton–Raphson iteration, after sir Isaac Newton and Joseph Raphson, an English mathematician who refined and simplified Newton's original method. Starting from an initial guess ($n = 0$), Newton iteration calculates an improved estimate of the root using the formula

$$x_{n+1} = x_n - \frac{f(x_n)}{f'(x_n)}, \tag{B.8}$$

where the prime indicates the first derivative of the function $f(x)$, the roots of which are being sought.

Newton's method usually converges rapidly, and it is much more likely to converge than the iterative method described in the previous section (Newton's method works for any sufficiently "well-behaved" function). It is also easy to refine the approximation to the solution to any desired degree, and the method can be extended to systems of equations (Newton–Raphson iteration is the solver of choice for many multiphase flow and transport models). The main drawbacks of the method are that it is somewhat more difficult to program than either trial and error or iteration (although not prohibitively difficult), and it requires careful consideration of the initial guess in order to converge to the desired root in cases when a function has more than one root. One additional difficulty is that, in order to find the first derivative of the function $f(x)$ and evaluate it at x_n, the user may need to resort to implicit differentiation (the technique of implicit differentiation is described in almost any standard calculus textbook). Despite these difficulties, which are generally slight, Newton iteration is probably the most commonly used root-finding method, and its many positive attributes far outweigh any minor difficulties in its application. You should most certainly learn how to use it and write a root-finding program of your own for future use.

References

Fairley, J.P. (2009) Modeling fluid flow in a heterogeneous, fault-controlled hydrothermal system. *Geofluids*, **9**, 153–166, doi:10.1111/j.1468–8123.2008.00236.x.

Logan, J.D. (2006) *A First Course in Differential Equations*, Springer Science+Business Media, LLC, New York.

APPENDIX C

Matrix solution for implicit algorithms

As explained in the text (Section 6.7.1), the explicit method for iteratively solving numerical approximations of finite difference equations suffers from numerical instability for timestep sizes greater than some value of the parameter $\Delta\tau/\Delta\xi^2$ ($\Delta\tau/\Delta\xi^2 > 0.5$ for one dimension, or $\Delta\tau/\Delta\xi^2 > 0.25$ for 2 dimensions). This restriction limits the usefulness of the explicit scheme, since the resultant timestep size (or coarse grid discretization) makes it difficult and computationally intensive to treat any but the simplest of problems.

To circumvent the difficulties of the explicit timestepping scheme, it is generally preferable to use implicit solution methods. Implicit methods rely on simultaneous solution of N unknowns with N relationships between them; in matrix notation,

$$[A][\theta] = [R], \tag{C.1}$$

where $[A]$ is the $N \times N$ coefficient matrix, $[\theta]$ is the $1 \times N$ matrix of unknowns, and $[R]$ is the $1 \times N$ matrix of boundary conditions and source terms.

There are many methods for solving equations of the type shown in C.1; however, most of the solution methods for the general problem are proprietary—that is, the aspiring modeler must purchase a library of equation solvers (e.g., LAPACK, PLAPACK, or similar), and then call the appropriate solver routine from within the user's own code. Although there is no special difficulty with this, the purchase of such libraries can be expensive.

C.1 Solution of 1D equations

For a particular special case of Equation C.1, a non-proprietary solver algorithm is available. The special case referred to is the *tridiagonal matrix*; that is, the matrix $[A]$ in which all entries are zero with the exception of the main diagonal and its two flanking diagonals. Tridiagonal matrices occur in 1D finite difference problems, and can be solved by the method known as the *Thomas algorithm*, named after the British physicist and applied mathematician Llewellyn Thomas [1903–1992]. The algorithm is based on the technique of Gaussian elimination,

Models and Modeling: An Introduction for Earth and Environmental Scientists, First Edition. Jerry P. Fairley.
© 2017 John Wiley & Sons, Ltd. Published 2017 by John Wiley & Sons, Ltd.
Companion website: www.wiley.com/go/Fairley/Models

which is taught in any basic course in linear algebra. The Thomas algorithm is a *method* for solving tridiagonal matrix equations, not a piece of computer code; however, many subroutines have been written to implement the Thomas algorithm. One such subroutine, written by me in Fortran 95, is included at the end of this appendix.

C.2 Solution for higher dimensional problems

The difficulty with using the Thomas algorithm in general is that it is limited to 1D problems, because problems in higher dimensions (2D and 3D dimensions) no longer yield tridiagonal matrices. However, a hybrid method of solution (i.e., an iterative matrix solution scheme) can be used to apply the Thomas algorithm to 2D and 3D problems. This is accomplished by breaking the problem into 1D strips, solving each strip independently using the Thomas algorithm, then breaking the problem into 1D strips in the orthogonal directions, re-solving the new 1D strips (using the solution from the previous iteration as the "initial guess"), and continuing to iterate in this fashion until the solutions from all the different orientations of the 1D strips converge to within tolerance. This method is sometimes known as *alternating direct implicit* (ADI) (Wang and Anderson, 1982). The actual implementation of this type of scheme is beyond the scope of this appendix; the interested reader is referred instead to Patankar (1980) and Wang and Anderson (1982).

C.3 The tridiagonal matrix routine TDMA

I give here a program listing, in Fortran 95, of my implementation of the Thomas algorithm.

```
SUBROUTINE TDMA(NX,A,B,C,D,T)
!
! subroutine TDMA.90 is a tridiagonal matrix solver
! (Thomas algorithm) written for use in a loop-level
! parallelized code (or not).
! NX is the number of unknowns (gridblocks minus boundaries)
! The coefficient matrices A,B,C,D, and T are defined
! as follows:
! A The coefficients of the lower diagonal
! B The coefficients of the main diagonal
! C The coefficients of the upper diagonal
! D The knowns; boundary conditions, source/sink terms, etc.
! T The solution matrix
```

```
!
! RECORD OF CHANGES ******************************
!
! 12.08.2009 JFairley Original code
! 02.27.2012 JFairley Added documentation about
! the arrays
!
! ************************************************
```

```
IMPLICIT NONE
```

```
! DATA DICTIONARY ******************************
!
INTEGER, INTENT(IN):: NX ! Number of gridpoints
REAL(KIND=8), INTENT(IN):: A(NX) ! 1D coeff matrix
REAL(KIND=8), INTENT(IN):: B(NX) ! 1D coeff matrix
REAL(KIND=8), INTENT(IN):: C(NX) ! 1D coeff matrix
REAL(KIND=8), INTENT(IN):: D(NX) ! 1D coeff matrix
REAL(KIND=8), INTENT(OUT):: T(NX) ! output temps
!
REAL(KIND=8) :: P(NX) ! Working values
REAL(KIND=8) :: Q(NX) ! Working values
REAL(KIND=8) :: denom ! intermediate result
INTEGER :: i ! loop index
!
! ************************************************
```

```
P(1) = B(1)/A(1) ! Put values in the first spots
Q(1) = D(1)/A(1) !
```

```
DO i = 2,NX ! This loop fills the matrices
denom = A(i)-C(i)*P(i-1)
P(i) = B(i)/denom
Q(i) = (C(i)*Q(i-1)+D(i))/denom
END DO
```

```
T(NX) = Q(NX) ! Starts the countdown of answers
DO i = NX-1,1,-1
T(i) = P(i)*T(i+1)+Q(i)
END DO
```

```
END SUBROUTINE TDMA
```

References

Patankar, S.V. (1980) *Numerical Heat Transfer and Fluid Flow*, Routledge, Taylor & Francis Group, New York.

Wang, H.F. and Anderson, M.P. (1982) *Introduction to Groundwater Modeling: Finite Difference and Finite Element Methods*, Academic Press, San Diego.

Index

accuracy
  and precision, 209
  defined, 209
ADE, 142–144, 157, 172, 173, 215, 220
  derivation, 146–148
  non-dimensionalization, 148–152
advection, 142, 152–155, 157, 164, 165, 170, 173
advection equation, 157–158, 160–164, 166, 169,
      172, 173
anisotropy, 76, 177–193
  coordinate axes orientation, 187–188
  defined, 177–178
  in finite differences, 189–191
  in transport, 169
approximation
  asymptotic series, 135
  using derivatives, 196
aspect ratio, 79, 82, 83
averaging
  between finite difference schemes, 117
  property of Laplace equation, 58, 82

Biot number, 71, 72, 166, 167, 237
boundary conditions, 69–72
  Cauchy, 165–167, 173, 220
  first kind, 48, 69–70, 72
  second kind, 69–70, 72, 74
    radial coordinates, 128
  third kind, 69–72, 166, 218
    confusion with Cauchy conditions,
      167
Boussinesq approximation, 120

Cauchy-Euler equation, 231
Cauchy-Riemann equations, 208
coefficient
  convective cooling, 41, 218
coefficients
  conductivity
    hydraulic, relationship to transmissivity,
      96
    hydraulic, tensor, 187–188
    thermal, 71, 220, 231
  diffusivity, 88, 98, 128

in similarity variable, 130
  in species transport, 142, 147–148, 167
  role in dimensionless time, 113–114
  thermal, 220
dispersivity
  hydrodynamic, 168
  longitudinal, 169
  mechanical, 168
  transverse, 169
permeability
  fault, 211
  relationship to conductivity, 215, 231
porosity
  in retardation, 168
  in transient flow, 93
  irreversible reduction of, 97
  role in pore velocity, 150
retardation
  defined, 168
storage, 95–98
  assumption of constant value, 97–98
  specific heat, 95
  specific storage, 95
  specific yield, 95–97
  storativity, 16, 95–96
  unconfined storativity, 97
tortuosity, 168
transmissivity, 76, 96, 127, 128
Courant number
  local, 171, 174
curvature
  defined, 51

Darcy's law
  analogy to Fick's law, 147
  analogy to Fourier's law, 218, 220
  differential form, 48, 59, 76, 112, 126, 127
  dimensionless, 21, 52, 112
  linearized, 15, 21, 52, 112, 226
  radial coordinates, 126, 127, 231
  substitute for q, 48, 59, 76, 94, 147, 216
  to calculate flux, 183
dimensionality
  reduction by approximation, 79–80
  reduction by symmetry, 67–68, 72–73

Models and Modeling: An Introduction for Earth and Environmental Scientists, First Edition. Jerry P. Fairley.
© 2017 John Wiley & Sons, Ltd. Published 2017 by John Wiley & Sons, Ltd.
Companion website: www.wiley.com/go/Fairley/Models

effective properties, 177–193
  averaging, 185–186
  conductivity, 186–187
    basin-scale, 14–16
    for layered systems, 181–184
    harmonic averaging for numerical simulation,
        190–191
    in fractured rock, 167
  diffusivity, 168–169
  dispersivity
    scale of measurement, 169
  exponential decay model, 14–16
  for layered systems, 181–184
  for uniform flow, 186–187
  fractured rock, 167
  of a flowpath, 227
  porosity
    defining an REV, 179–180
  relationship to REV, 181
error
  relative, 200–202
expected value, 184–185
  of a continuous distribution, 184
  of a discrete distribution, 184–185
exponential decay, 9–10, 106, 114, 138, 143
  alternative names, 10
  described conceptually, 9
  model, 29, 30, 36, 113–114
    calibration, 21, 113
    calibration, curve matching, 23
    calibration, semi-log plots, 22–23
    derivation, 15–16
    non-dimensionalization, 17–18
    two domains, 24–26
exponential integral, 133, 134, 139
  asymptotic expansion for large u, 135
  Taylor series expansion, 134–135

Fick's law, 147, 148
finite differences
  ADE, 169
  applied to Laplace equation, 54
  applied to Poisson equation, 81
  direct methods, 56–57
  first derivative, 27–28, 54
  heterogeneous properties, 189–192
  in 2-D, 81, 83–84
  in 3-D, 82
  in transport, 169
  integrated, 143, 174
  iterative methods, 55–56
    Gauss-Seidel, 56
    Jacobi, 55–56
  second derivative, 54
    in 2- and 3-D, 81–82
  transient systems, 114
    Crank-Nicolson, 117–118
    explicit schemes, 115
    implicit schemes, 115–117
  tridiagonal matrices, 57
  with source term, 81–84

finite elements, 171–172
Fourier
  Fourier's law, 218, 220, 231
  historical figure, 107
  number, 113–114, 164, 173, 237
  series, 101, 107–109, 113, 120, 176
  transforms, 101
Froude number, 162

heterogeneity, 76, 177–193, 212
  defined, 177
  effect on potentiometric surface, 53
  in finite differences, 189–191
  in layered systems, 182
  in transport, 167, 169
homogeneity, 12
  assumption of, 44, 66, 179–180
  defined, 177

initial conditions, 16, 40, 41
  dimensionless, 18
  for similarity variables, 131
isotropy, 12
  defined, 177–178
  of correlation structure, 186

Laplace
  equation, 7, 42–43, 51, 57–58, 62, 63, 66, 70, 78,
      82, 85, 87, 118, 141, 143
    applications in the real world, 53–54
    assumptions, 53
    averaging property, 58, 82
    derivation, 47–48
    non-dimensionalization, 48–49
    zero curvature prediction, 52
  historical figure, 42, 107
  transforms, 101

Mach number, 162
mean
  arithmetic, defined, 182
  geometric, defined, 185
  harmonic, defined, 184
model
  sensitivity, 201–206
    and robustness, 202, 205
    applications, 204–205
    defined, 201–203
    examples, 203–204, 227–229
    sensitivity studies, 205–206
    to small errors, 203

non-homogeneous
  boundary conditions, 236
  governing equation, 62, 63, 101, 222
nondimensionalization
  Darcy's law, 52
  equivalence between head and discharge, 21
  heuristic approach, 236–237

identifying non-unique parameter values by, 19
of parameters, 35–36, 78
of the dependent variable, 17, 34–35
of the independent variable, 17–18, 49
similarity, 130–132
nonhomogeneous
  boundary conditions, 72

Ockham's Razor, 13, 14

Peclet number, 151–155, 157, 162–165, 167, 170,
      173, 175, 221, 222, 225, 226, 236, 237
  local, 171, 173, 174
Poisson equation, 7, 62–63, 84, 87, 118, 141, 143,
    218
  curvature predicted by, 78–79, 87
  derivation, 74–76
  non-dimensionalization, 76–79
precision
  and accuracy, 209
  defined, 209

relative error, 200–202
REV
  defined, 180
robustness of models, 202, 205
rules
  for model development, 5–7

salt tank model, 31, 39
  behavior in the limits, 37–38

derivation, 33–34
non-dimensionalization, 34–36
storage
  geologic water storage, 30
  in soils, 24
  Inka, 30
  sources of water from, 95
  water released from, 16

Taylor series
  use in approximation, 197–198
TDS, 32–35, 37–40
Theis equation, 122, 134, 139–140, 218, 237
  conceptual model, 125
  control volume, 126
  derivation, 125–128
  non-dimensionalization, 128–132
  non-ideal behavior, 138–139
transient diffusion equation, 7, 87–88, 118,
      139–141, 143, 144, 148, 170, 172
  derivation, 92–94, 98
  non-dimensionalization, 99–100

uncertainty
  of data, 199
  propagation in models, 200–201

variance
  of a continuous distribution, 185
  of a discrete distribution, 185